SEX TOURISM IN AFRICA

New Directions in Tourism Analysis

Series Editor: Dimitri Ioannides, Missouri State University, USA

Although tourism is becoming increasingly popular as both a taught subject and an area for empirical investigation, the theoretical underpinnings of many approaches have tended to be eclectic and somewhat underdeveloped. However, recent developments indicate that the field of tourism studies is beginning to develop in a more theoretically informed manner, but this has not yet been matched by current publications.

The aim of this series is to fill this gap with high quality monographs or edited collections that seek to develop tourism analysis at both theoretical and substantive levels using approaches which are broadly derived from allied social science disciplines such as Sociology, Social Anthropology, Human and Social Geography, and Cultural Studies. As tourism studies covers a wide range of activities and sub fields, certain areas such as Hospitality Management and Business, which are already well provided for, would be excluded. The series will therefore fill a gap in the current overall pattern of publication.

Suggested themes to be covered by the series, either singly or in combination, include—consumption; cultural change; development; gender; globalisation; political economy; social theory; sustainability.

Also in the series

Cultures of Mass Tourism
Doing the Mediterranean in the Age of Banal Mobilities
Edited by Pau Obrador Pons, Mike Crang and Penny Travlou
ISBN 978-0-7546-7213-5

The Framed World
Tourism, Tourists and Photography
Edited by Mike Robinson and David Picard
ISBN 978-0-7546-7368-2

Brand New Ireland?
Tourism, Development and National Identity in the Irish Republic
Michael Clancy
ISBN 978-0-7546-7631-7

Cultural Tourism and Sustainable Local Development
Edited by Luigi Fusco Girard and Peter Nijkamp
ISBN 978-0-7546-7391-0

Sex Tourism in Africa
Kenya's Booming Industry

WANJOHI KIBICHO
Moi University, Kenya

ASHGATE

© Wanjohi Kibicho 2009

All rights reserved. No part of this publication may be reproduced, stored in a retrieval system or transmitted in any form or by any means, electronic, mechanical, photocopying, recording or otherwise without the prior permission of the publisher.

Wanjohi Kibicho has asserted his right under the Copyright, Designs and Patents Act, 1988, to be identified as the author of this work.

Published by
Ashgate Publishing Limited
Wey Court East
Union Road
Farnham
Surrey, GU9 7PT
England

Ashgate Publishing Company
Suite 420
101 Cherry Street
Burlington
VT 05401-4405
USA

www.ashgate.com

British Library Cataloguing in Publication Data
Kibicho, Wanjohi.
   Sex tourism in Africa : Kenya's booming industry. -- (New directions in tourism analysis)
   1. Sex tourism--Kenya. 2. Tourism--Kenya--History. 3. Sex-oriented businesses--Kenya. 4. Kenya--Social conditions--1963-
   I. Title II. Series
   306.7'4'096762-dc22

Library of Congress Cataloging-in-Publication Data
Kibicho, Wanjohi.
   Sex tourism in Africa : Kenya's booming industry / by Wanjohi Kibicho.
      p. cm. -- (New directions in tourism analysis)
   Includes bibliographical references and index.
   ISBN 978-0-7546-7460-3 (hardback) -- ISBN 978-0-7546-9844-9 (ebook)
1. Sex tourism--Kenya. 2. Prostitution--Kenya. I. Title.

HQ260.5.K53 2009
306.74096762--dc22

2009021838

ISBN   9780754674603 (hbk)
ISBN   9780754698449 (ebk)

Mixed Sources
Product group from well-managed forests and other controlled sources
www.fsc.org Cert no. SA-COC-1565
© 1996 Forest Stewardship Council

Printed and bound in Great Britain by
MPG Books Group, UK

Contents

List of Figures		*vii*
List of Tables		*ix*
Acknowledgements		*xi*
List of Abbreviations and Acronyms		*xiii*
Foreword by Professor Chris Ryan		*xv*
Setting the Scene		1
1	Conceptual Background	13
2	Sex Trade and the Law: An Epitome of Legal Ambiguities	41
3	Tourism Development in Kenya	57
4	Sex Tourism in Kenya	79
5	Kenya's Tourism Industry: A Facilitator of Romance and Sex	97
6	Socio-economic Base of Sex Tourism in Kenya's Coastal Region	121
7	Sex Trade in Kenya's Coastal Region: A Sword that Cuts with Both Edges	143
8	Sex Trade in Malindi: Roles Male Sex Workers Play	157
9	Tourism's Influences on the Sex Workers' Operations in Malindi	179
General Conclusion		207
Epilogue		*215*
Bibliography		*217*
Index		*227*

To my late brother Hillary Githaka KIBICHO
(who was brutally murdered during the final phase of this project)

To laugh often and much;
to win the respect of intelligent people and
the affection of children;
to earn the appreciation of honest critics and
endure the betrayal of false friends;
to appreciate beauty;
to find the best in others;
to leave the world a bit better,
whether by a healthy child,
a garden patch or
a redeemed social condition;
to know even one life has breathed easier
because you have lived.
This is to have succeeded!

(Ralph Waldo Emerson: 1803–1883)

List of Figures

1.1	Sex trade spectrum	32
1.2	Conceptual model of transactions between members of a sex trade community	34
3.1	Location map of Kenya	58
3.2	Evolution of tourist arrivals in Kenya (1978–2005)	60
3.3	Tourism destination life cycle	62
3.4	Factors influencing Kenya's sex tourism growth – exploration through consolidation phases	76
4.1	Principal types of sex trade in Kenya	91
6.1	Kenya's coastal region and the area of the case study	122
6.2	Categorisation of CSWs in Kenya's coastal region	139
9.1	Types of sex trade based on zones of operation in Malindi	188
9.2	Positive–negative continuum of tourism-related sex trade	196
10.1	Tourism-oriented sex trade in Kenya	208

List of Tables

3.1	National earnings from leading export crops and tourism (in K£ billions)	61
3.2	Hotel bed occupancy (%)	73
5.1	Origin of female sex tourists in Malindi ($n = 68$)	102
5.2	Age groups of female sex tourists in Malindi ($n = 68$)	102
6.1	Hotels closed September 1997–March 1998	123
6.2	Factors leading Kenyans into the sex business	127
7.1	Risks encountered by sex workers ($n = 183$)	144
7.2	Responses regarding condom usage and their role in reducing the spread of STDs ($n = 183$)	145
8.1	Gross Domestic Product by region – Malindi District	159
8.2	Monthly pay by occupation in Malindi Town, 2005	160
8.3	Survey items	165
8.4	Percentage distribution for the perceived linkage between tourism and the sex trade ($n = 73$)	166
9.1	HIV-AIDS prevalence by division in Malindi District, 2002	182
9.2	HIV-AIDS prevalence by industry, 2002	182
9.3	Price list for commercial sexual services	190

Acknowledgements

It is obvious that the ideas, arguments and examples scattered through every page did not just occur to me as I wrote this book. They are the product of reading, talking, corresponding, and often arguing with friends, colleagues, students and authors in many countries over the past decade. It has been produced with various forms of assistance, encouragement and support from different individuals and organisations. Thus I gratefully acknowledge the diverse practical and emotional support that I received from various quarters. Professors Bernard Lane and Bill Bramwell (co-editors of the *Journal of Sustainable Tourism*) first initiated me into the mysteries of scientific writing: they kept reminding me 'do not get discouraged by the unsparing comments by the reviewers'. The illustrations in the book have been produced with the assistance and expertise of several people. Thanks are offered in particular to Professor Laure Charleux (Université Lyon 2), who introduced me to the world of cartography.

I am grateful to Professor Erik Cohen, currently a professor of sociology at the Hebrew University, for his enthusiasm and help with this work. Without his aid, this book would have been much harder to write. His professionalism has been appreciated. Professors Chris Ryan (The University of Waikato) and Suzanne LaFont (The City University of New York)'s influence on the way I look at tourism and its effects, especially to the host communities, cannot go unmentioned. I am grateful to various other colleagues for useful discussions and comments, and I especially acknowledge the input and criticism of Professors Altaf Sovani and Paula Kerr (both of Algonquin College – Ottawa, Canada) which resulted, I think, in a better contribution to this modest book. A special mention goes to late Wahome Mutahi (*Whispers*), whose belief in my abilities never wavered and who always believed in the relevance of this study.

Some of the most important people involved in the compilation of this book I have yet to meet. They are the people to whom I sent questionnaires or enquiring letters. I am grateful for the trouble they have taken, and hope they will keep in contact so that I may remain up to date on their activities. In addition, due acknowledgement is made to various other sex workers – courageous and strong Kenyans coping under conditions of disadvantage with diminishing government tolerance. In relation to the foregoing, it is necessary to note that all sex workers' and sex tourists' names used in this text are fictitious. However, the interviewees themselves proposed these pseudonyms. They are their trade names.

I am indebted to the staff of Ashgate Publishing, whose patience, comments and advice have proven invaluable at all stages in the compilation of this book. More precisely, many thanks go to Valerie Rose (Senior Commissioning Editor) and Professor Dimitri Ioannides (editor of the *New Directions in Tourism Analysis* series), who perceived the germ of a book in my first draft, provided me with helpful readers' reports, and has shown a commitment to producing a book that reflects the messages of the written text. Profuse thanks to Sarah Horsley and Pam Bertram for their timely responses to my unending enquiries. Detailed critiques of an earlier draft of the manuscript by an anonymous reviewer proved extremely helpful.

Finally, I owe special thanks to my family (Nathalie Charun, Karen Sephora Kibicho, Wesley Naftie Kibicho, David K. Kibicho Jnr, Ann N. Kabugua, Peninnah W. Kabugua, Edna W. Kabugua, Sephora Kibicho) and friends (Dr. Munene Mwaniki, Kariuki Gichuki, Michael Wanjau, Elijah Kasati, Daniel Kuting'ala, Kariuki Ndune, Jean-Michel Dewailly), who have tolerated interminable references to 'the sex book' for the last 24 plus months!

Although the material for *Sex Tourism in Africa: Kenya's Booming Industry* is new, some chapters develop themes of articles that have been previously published. I therefore thank the Permission Managers of Multilingual Matters (*Journal of Sustainable Tourism*) and Cognizant Communication Corporation (*Tourism Review International*) for kindly allowing me to reprint (in revised form) Chapters 6 and 7 respectively, and the Managing Editor of the *Annals of Leisure Research* for permission to reprint Chapter 9. Every effort has been made to obtain permission to reproduce material contained in this book. If proper acknowledgement has not been made, I would like to take this opportunity to apologise to any copyright holders whose rights I may inadvertently infringed.

Although each of the above people, both advertently and inadvertently, has contributed to the book, I alone must be held responsible for any remaining shortcomings.

Wanjohi Kibicho, PhD
Ottawa, Canada
August 2009

List of Abbreviations and Acronyms

This glossary identifies all of the common, and most of the less common, acronyms and initialisms used in this book. Many of these entries have been excluded from the index for the sake of brevity, although those that occur frequently in the text, or in the opinion of the author are important, may be found in the index.

AICE	American Immigration and Customs Enforcement
AIDS	Acquired Immune Deficiency Syndrome
AT&H	African Tours and Hotels
AWF	African Wildlife Fund
CSW	commercial sex worker
GDP	Gross Domestic Product
HIV	Human Immunodeficiency Virus
IFC	International Finance Corporation
ILO	International Labour Organisation
IMF	International Monetary Fund
Interpol	International Police
JICA	Japanese International Co-operation Agency
K£	Kenya pound
KTB	Kenya Tourist Board
KTDC	Kenya Tourism Development Corporation
KWS	Kenya Wildlife Service
MCSW	male commercial sex worker
PA	Protected Area
RLD	red light district
SAP	Structural Adjustment Programme
SARS	Severe Acute Respiratory Syndrome
STC	sex trade community
STD	sexually transmitted disease
TFUI	touch-and-feel user interface
UNICEF	United Nations Children's Fund
VRML	Virtual Reality Modelling Language
WCMD	Wildlife Conservation and Management Department
WTO	World Tourism Organisation
WWF	World Wide Fund

Foreword

Sex tourism is a complex, multi-faceted activity of so many hues and colours, that whatever one's beliefs about this human activity, the phenomenon will provide sufficient examples to reinforce any given prejudice. As academic researchers, those located in universities have a responsibility to illuminate *all* aspects of this activity. It is true that there exists human trafficking, a feminisation of poverty that means that some females feel they have little choice but to turn to sex work, but equally it is also true that many sex workers forcibly reject any notion of victimisation while others have argued that sex work can lead to personal senses of emancipation and certainly economic independence. Within any given context, at a given place and time, individuals involved in sex tourism, and those responsible for the provision of aid to sex workers and the enforcement of legal requirements – all make their own decisions – some do so to mitigate problems and provide support, and others in ways that degrade, criminalise and abuse the sex workers and his or her clients. The very fact that academic researchers can act as critical agents of change means they challenge the perceptions of both those who condemn and those who support sex work – for good research will give a voice to the oppressed and the emancipated, recognising the legitimacy of both ends of a spectrum, and the confusions and complexities that lie betwixt and between. It will also mean that good research leads to conclusions that are often themselves contentious because they challenge the preconceptions of at least one part of society.

Given (a) the fundamental nature of sexual urges, (b) the complex ways in which societies have sought to manage sex work, (c) the ethos of tourism as a period of social irresponsibility away from the constraints of 'home', (d) the personal consequences of decisions that subsequently arise from engaging in commercial sexual acts, and (e) the nature of sex tourism and the wider social, economic and medical consequences that emanate from it, it is not surprising that sex work has engaged the minds of researchers and a curiosity on part of wider audiences. Such research may be motivated by different concerns – clinical, social, economic – but each researcher will end questioning not only wider social values but often also their own. There is a rich literature on sex tourism, but much of this has concentrated on three parts of the globe – namely Europe, North America and the Caribbean, and also Asia. Wanjohi Kibicho has added to our understandings of Kenyan sex tourism by complementing and supplementing past work such as that of Rachel Spronk's *Ambiguous Pleasures* (which reports research undertaken in Nairobi) by providing an analysis of sex work along the Mombasa coastline of Kenya and elsewhere in that country. Kibicho provides a wider canvass than Spronk's seminal work by giving voice not only to the sex workers and their clients, but by also incorporating data and quotations from individual members of the police force and those charged with the enforcement of laws that date from past colonial regimes that reflected a British Victorian view that sex work was the by-product of dysfunctional women. He seeks to draw out the linkages between tourism as a basis of demand for sexual services, and the social context that provides these services by both males and females. Thus he provides a voice for both client and sex worker, revealing individual comments that illustrate stereotypes on both sides, and the complexities that exist simultaneously with those stereotypical views. He draws our attention to the work of bodies such as ECPAT, and the Malindi Welfare Association. He notes the different infrastructures of society, law enforcement, hotel construction, tourist flows and diseases such as AIDS/HIV.

Sex Tourism in Africa: Kenya's Booming Industry ends by concluding that the economic gains from sex tourism for Kenya are significant, and raises a question as to whether sex workers are 'a marginal layer in the social fabric' or 'a creative and innovative force' that is rightly entrepreneurial. His final words are in the form of a challenge – should Kenya consider the decriminalisation of prostitution, just as been the case of New Zealand where I sit writing these words. My own observation is that in New Zealand decriminalisation has not led to the undermining of society and arguably there exist many worse influences such as the ease of access to alcohol that feeds violence and certainly more loss of life than occurs by visiting a sex worker. By raising the question in the context of Kenya, Kibicho reinforces the notion that sex work in Kenya, as elsewhere, is not without benefits, but needs to be normalised whereby sex workers can claim the protection of the police without fear, and where the dangers of organised crime can also be minimised. But such 'normalisation' challenges the views of many in society, and so Kibicho's question is indeed fundamental in its nature. However, in this book he has marshalled many voices that generate a legitimate basis for such a challenge, and thereby has performed a service for many.

Professor Chris Ryan
University of Waikato Management School
Hamilton, New Zealand
August 2009

Setting the Scene

Background

The pleasures of sun, sea, sand, safari[1] and sex provide a major incentive for tourists to travel to African destinations (Kenya 1967, 1987, 1998, 2006; De Kadt 1979; Bachmann 1988; Sinclair 1990; Dieke 1991; KWS 1995, 1997; Kibicho 2005b, 2007). Consequently, about 3 per cent of world tourism activities take place in Africa as international travellers spend their holidays in tourism destinations in the continent (WTO 2004). A holiday is a form of non-work activity with particular spatial and temporal dimensions – it takes place away from home and over a prolonged period. Importantly, what occurs during that time could well be similar to non-work activity when not on holiday. Consequently, the connection between sex and tourism should not come as a surprise. Sex is natural in human life. If people engage in sexual activities at home, then it is only logical for one to expect them to be involved in sex while in a visited destination. As Bauer and McKercher (2003: 4) note, 'tourism only provides another setting for sexual activities. It only exposes the creative impulse of the sexual drive.' Moreover, 'in most cases the relationship between sex trade and tourism is positive, involving consenting adults engaging in a mutually gratifying, and, often, relationship-reinforcing activity' (Jago 2003: 134). However, in many situations, this linkage (sex trade–tourism) is neither neat nor pretty, but complex and difficult to describe and understand.

While sex trade, a polite term for prostitution, can generally be defined as emotionally neutral, indiscriminate and specifically remunerated sexual services (see Chapter 1), sex tourism is characterised by tourism whose principal motivation is consumption of commercial sexual relations (Graburn 1983; Cohen 1988a, 1989, 1993; Leheny 1995; Pruitt and LaFont 1995). In other words, the main components of sex tourism are travel, sex and financial exchange (Günther 1998). Thus, intent is an important factor in describing sex tourists, and indeed sex tourism in general (see Chapters 1 and 2).

---

1   *Safari* is a Kiswahili (Kenya's national language) word meaning 'to unveil' or 'to enter upon a journey or an expedition'. Its present-day context was introduced by the British, who undertook hunting safaris/expeditions in Kenya, then known as British East Africa, in the late 1800s (see Chapter 3).

Tourism-related Sex Trade

As mentioned elsewhere, tourism development leads to several impacts on a destination. The common trends in past works dealing with this subject of tourism impacts have been to focus on the environment, economic and socio-cultural impacts from a rather general approach. A lot has been written and said about the linkage between tourism and sex trade (Fanon 1966; De Kadt 1979; Mingmong 1981; Cohen 1982, 1988a, 1988b, 1989, 1993; Graburn 1983; Archavanitkul and Guest 1994; Philip and Dann 1998; Ryan and Hall 2001; ECPAT 2002; Afrol 2003; Kibicho 2004a, 2004b, 2005b). There are those studies that intend to reveal the male sex tourist flow from the developed to the developing countries with all its associated effects (O'Grady 1992; Launer 1993; Ackermann and Filter 1994). Falling in the same category, both conceptually and geographically, are academic studies that cover the relationship between tourism and sex trade in, predominantly, Southeast and East Asia (Hall 1994, Leheny 1995; ECPAT 2002; Jallow 2004). Strangely, African destinations are generally ignored in these studies, but they were highlighted by Sindiga (1995) and mentioned by Ndune (1996). However, in the recent past, a number of researchers have shown interest in studying the effects of sex tourism in Africa (ECPAT 2002; Sarpong 2002; Afrol 2003; Jallow 2004; Kibicho 2004a, 2004b, 2005b). According to ECPAT (2002), as a result of the restrictions that are being developed in Asia, combined with lack of controls in the African countries, the continent has become a favourite destination for tourists who have previously gone to Asia for sexual services. As with other tourism-related impacts in most of the African tourism destinations, however, this issue cannot be tackled effectively due to the absence of 'hard' data (Afrol 2003; Jallow 2004).

According to Sindiga (1995) and Ndune (1996), African society accepts that sex trade, which has been in existence for a long time, is a social problem. To them it means that there are some social processes which are failing to function. With this in mind, therefore, a study of such malfunctioning as a short cut to its understanding is long overdue. However, as a touchstone for such a study, a (potential) researcher needs to appreciate the fact that there are many social processes, of greater and lesser drives, of shifting strengths, and inter-operative factors. Objectively, a social process is a series of social changes. Structurally, it is the mode which the social changes of a given series follow. Subjectively, a dynamic social process is found in the changes in attitudes and values of those persons who figure in any series of social changes. Intrinsically, a social process is a dynamic, moving equilibrium of human energy. Consequently, sex trade, and indeed its relationship with tourism, should be evaluated from a more conceptual societal macro-perspective.

Moreover, Kibicho (2005b: 258) warns: 'no one can be expected to offer diagnosis of social problems that will be of value in a problem-sore world unless he is first willing to locate the monkey wrenches in the processual machinery', thus further validating the importance of a detailed treatment of the current subject. The pitfall in discussing (tourism-related) sex trade is that we maintain an illusion of knowing what the 'solution/problem' is when we really do not. When this approach is adopted, the systemic groundings of sex trade are likely to be ignored, as an endemic social challenge is individualised and depoliticised. The promise is that we will be better off in the future by examining the concept of sex tourism than by not examining it. From an academic point of view, therefore, there is no need for the authorities concerned to spend time and other resources (which are in short supply) organising for periodic police crackdowns on commercial sex workers (CSWs) rather than being directed against the conditions which generate sex trade. It is dealing with the symptoms of the problem rather than the predisposing factors – roots – and how the problem could be generally addressed (Cohen 1982; Ndune 1996). This book addresses these root causes in an attempt to go beyond the fallacies of most of the previous authors and tourism authorities.

Methodological Critique

Like many studies dealing with the current subject, there were many methodological, definitional and attitudinal difficulties in this research. First, discussions of the relationship between tourism and sex trade are frequently speculative, while studies of the subject often become an act of sensationalism and/or moralism (Cohen 1982, 1988a; Ryan and Hall 2001). They frequently suffer from cultural blindness and sexual taboos which affect attitudes towards sex trade. It has always been a sensitive subject, hence the triviality of study as well as pursuit (Cohen 1982; Ndune 1996). Consequently, information on sex trade is often restricted to approximation of the numbers of CSWs in a destination. It is further limited by a tendency to condense complex processes into simple descriptions, to avoid analysis and criticism. This work therefore tries to bridge the research gap by providing a theoretically informed analysis of the tourism-oriented sex trade in Kenya. Like many other studies on the subject of sex trade, however, this book is expected to invite both specific and general criticisms. This is more so due to the sensitivity with which the word 'sex' is treated by many societies (Backwesegha 1982). Maybe this is why Cohen (1982: 406) feels that 'the issue is extremely touchy'. There is a certain uneasiness in speaking about the fulfilment of sexual appetites, because it sometimes implicates the researcher in legitimising sexual practices. This accounts for the general absence of systematic and in-depth studies on the subject. Depending on personal prejudices, and indeed unwillingness to accept the reality, therefore, it is not expected that all readers will share the conclusions reached in this text.

Second, data on the current topic are highly deficient. This is largely due to the illegal and informal nature of the sex trade in Kenya, not to mention the social stigma attached to CSWs in general (see Chapter 7). These high levels of data deficiency made it extremely difficult for this researcher to conduct a prior review of the extent and magnitude of the relationship between tourism and sex trade in the country.

Apart from the scarcity of reference materials, identification of CSWs in Kenya, and in the specific resorts in particular, was the most difficult task of the current study. First, sex trade in most of Kenya's tourism resorts is a 'nocturnal profession', thus making it difficult for the researcher to meet with CSWs during the day. Then, as noted earlier, due to its illegality, there is a complete lack of official documentation of the trade, which makes it impossible to generate a sampling frame. However, with the help of a group of CSWs from Malindi resort, this researcher managed to arrive at an approximate number of sex workers, especially in the tourism resorts at the coastal region. As indicated in Chapter 9, CSWs in Malindi resort are highly organised to extent of having an 'unofficial' social welfare organisation (Malindi Welfare Association – MWA) with rough details on sex trade in the region. These details are more exact for members of the MWA, but not very specific about non-members. Nevertheless, the MWA's secretary can give the number of CSWs in Malindi resort with certainty. This is possible as the MWA is always informed, by its members and non-members, of new entrants to the profession as well as those leaving the profession. Due to lack of such organisation of CSWs in the rest of the country, and other coastal region resorts in particular, this researcher had to use the MWA Secretary, in return for a fee, to determine respondents in the seven resorts identified in Chapter 6. Being an insider of the profession, this task seemed not very difficult as:

1. she knew the relevant persons (sex workers) to approach for the interviews;
2. the identified CSWs divulged information with ease as they had confidence in her, unlike when the author approached them alone.

As noted in Chapter 6, CSWs in Kenya's coastal region have relatively fixed operation areas. This made it easy for the researcher to establish the approximate number of CSWs operating in a given area within a resort. It should be noted that CSWs operating in the same area know one another very well (Kibicho 2004a, 2004b, 2005b). Based on these sub-totals, a total of CSWs for the resort was then computed. This was repeated for each resort, and then the grand total for Kenya's coastal region was derived. However, this is not to suggest that the number of CSWs used in this study is the total number of sex workers working in the region. It may be lower or higher than this: lower due to the few cases of CSWs with multiple areas of operation, who thus could have been counted twice or thrice, and higher due to seasonal sex workers.

All primary data presented in this book, unless noted otherwise, are based on several surveys and the author's observations in the field from June 1997 to March 2006. An in-depth questionnaire survey in Kenya's coastal region was conducted among CSWs from January to July 2002. Seven major tourism resorts in the region were used as the study units: Bamburi-Kisauni, Diani-Ukunda, Malindi, Mombasa Island, Mtwapa, Shimoni and Watamu (see Figure 6.1). In addition, the author conducted in-depth structured interviews with CSWs, sex tourists – both male and female – and various tourism stakeholders in Kenya.

Owing to difficulties associated with investigating sex trade using 'traditional' research methods, this study employed a combination of research approaches, both quantitative and qualitative. Thus, field data were obtained through informal and structured interviews with diverse stakeholders in the different study areas. Life history and ethnography research methodologies were heavily used in this study. Life history methodology assumes that the respondents are telling the truth about their lives in so far as they understood and remembered events. The technique of revisiting and reinterpreting the material in subsequent interviews ensures that the stories are consistent (see, for example, Middleton 1993). On the other hand, ethnography is the art and science of describing a group or culture. The objectives of exploring, investigating, studying and reporting a social phenomenon like sex trade thus match the scope of this study method.

Relevant life histories of a number of informants were tape-recorded. Interviews were conducted in English, French and Kiswahili. In most cases, the author participated in a number of tourism-related sex trade activities, but these did not include (commercial) sexual liaisons with the sex workers. These activities, like going to discothèques, visiting massage parlours and taking up temporary residence in local accommodation established a perfect arena for the examination of the interaction between CSWs and their clients. Data obtained through participant observation were supplemented by interviews with a number of Kenya's tourism stakeholders, especially for information presented in Chapter 3. On the other hand, secondary data were obtained from both government and non-governmental institutions. The collected data were analysed, where applicable, with Statistical Package for Social Scientists (SPSS) (see, for example, Chapters 6–8).

Purpose

Few policy issues have generated greater interest and controversy than how to deal with sex trade in general, and sex tourism in particular (Cohen 1982; Truong 1983; Kempadoo 1999; Ryan and Hall 2001). The debate over how to do this is intense and often polemical. A central part of this debate revolves around whether to legalise the trade or not (Smart 1992). Surprisingly little compromise has emerged about the issue, with advocates of one approach or another continuing to argue for their preferred 'solution', while policymakers continue to make important decisions about the trade (Caplan 1984; Ackermann and Filter 1994; Pruitt and LaFont 1995; Cabezas 1999; Brown 2000).

This book offers a different perspective when dealing with the issue of tourism-oriented sex trade in Africa. In doing so, the author builds upon past research published elsewhere (Kibicho 2003, 2004a, 2004b, 2005a, 2005b). Unlike some previous large comparative studies that aim at proving that one approach is better than another, the text examines different factors that affect the operations of participants at an extreme micro-level within (sex) tourism destinations. Rather than viewing these destinations as 'black boxes' and attempting to evaluate their 'performances/outputs' under different conditions, this text dares to ask what goes on with individual CSWs. It explores how different (tourism) policy initiatives influence the key actors in what I will call the sex trade community (STC – including CSWs, sex tourists and sex trade organisations) and the interactions between them (Figure 1.2). Even though these actors are recognised as important (see, for example, Naibavu and Schutz 1974; Pruitt and LaFont 1995; Ndune 1996; Muroi and Sasaki 1997; Cabezas 1999; Ryan and Hall 2001; Trovato 2004; Kibicho 2005b), we do not have an explanatory model of how they interact. The aim of this book and the theory underlying it is to provide such a model and apply it to questions about the CSWs' operations in Kenya. In so doing, it bypasses a number of major roadblocks to understanding the organisation of CSWs in Africa. It develops an innovative theoretical framework, drawing on a body of social, planning, business, environmental and economic thoughts, that help us to understand why and how different approaches to control sex trade do or do not work.

In this book, sex tourism is viewed as a crucial agent of socio-economic change. It alters the social fabric of tourism destinations, while much of its development pace and direction are guided by broader global forces. As a consequence, (sex) tourism, and its high visibility, is often viewed as the primary cause of socio-economic change rather than as one of many agents (see, for example, Graburn 1983; Truong 1983; Cohen 1988a, 1988b; Ackermann and Filter 1994; Harrison 1994; Leheny 1995; Kempadoo 1999; Ryan and Hall 2001). Within this framework, the book is committed to finding ways in which CSWs can survive as social units while engaging in a range of tourism-related commercial endeavours that enable them to meet their individual goals and aspirations. To achieve this objective, it evaluates the multi-dimensional nature of (sex) tourism and its significant social and economic effects. These features are underpinned by the premise that sex tourism is an economic undertaking, and to properly understand an STC's prospects and impacts, the whole activity should be examined in a social context. To do this the book strives to:

1. move towards an objective appraisal of the relationship between tourism and the sex trade;
2. evaluate each in terms of their articulation with and impact on poverty;
3. demonstrate the existence of diverse and discrete sectors within the sex industry, which articulate with the tourism industry in different ways;
4. shift the debate from a focus on moral issues to an investigation of the effectiveness of entrepreneurial strategies, and the promotion of pro-poor development strategies as a way of combating the most exploitative manifestations of sex tourism.

As a result, the ways social and economic sciences' interests overlap in sex tourism issues are demonstrated throughout the text.

With regards to the unceasing expansion of sex trade in Kenya's key tourism destinations, one of the most important questions is: What is the role of the tourism industry in the growth of sex trade in Kenya? Specifically, how does tourism affect the environment in which the key sex actors carry out their respective roles as sexual service providers and consumers? How do socio-economic needs influence transactions between members of the STC? These are some of the questions this book attempts to address. Consequently, I have opted to discuss a wide range of sex tourism-related topics and to provide a general theoretical framework for analysing them – either separately or in combination. Because of the scope of coverage attempted here, it has not been possible to explore all the topics in the greatest depth or detail. I have, necessarily, been selective rather than encyclopaedic – with respect to research data, interpretations and even the choice of topics themselves.

Overall, therefore, the goal of this book is to examine some of the features of tourism and sex trade that currently exist in Kenya. However, it does not attempt to examine the morality of sex trade or who is to blame for the 'oldest profession' in the world. It only tries to contextualise this trade in relation to Kenya's tourism development. Thus, it seeks to provide an authoritative reference for teachers, students, and other persons who need to know more about sex tourism in Africa. But it bears stressing that some arguments made in this book may be uncongenial for those with little appetite for pointed analysis of a 'sensitive' subject like tourism-related sex trade. At the same time, those with a genuine interest in venturing beyond established orthodoxies and moralistic and simplistic approaches to the contentious sex trade phenomenon will find the text to be useful and worthy of close attention. With this in mind, let me bring this section on purpose of this book to a close by stressing that good scholarship, in the form of reasonable substantiation of suppositions and sourcing of information, overrides sensationalism, propaganda and egocentrism in this modest text.

The primary audience for this book is senior undergraduate and postgraduate students of tourism. As the book approaches sex tourism from an interdisciplinary perspective, it is anticipated that sociologists, psychologists, health practitioners, policymakers and politicians will also find it a useful tool.

Structure of the Book

To enable a systematic and exhaustive investigation of the nexus between sex trade and tourism in Kenya, the text begins with a general evaluation of the development of the national tourism industry and the organisation of sex trade in the country. It then narrows its scope to Kenya's coastal region before shifting its attention to the Malindi Case Study in the last two chapters. The choice of these two study areas (the coastal region and Malindi) is based on the fact that Kenya's tourism industry is heavily oriented towards its coastal region, with coastal tourism contributing to over 60 per cent of national tourism activities (see Chapters 3 and 6). Thus, it is assumed that the manifestation of the sex trade–tourism linkage in these study areas is a miniature replica of the national sex tourism story – the masked face of a booming industry.

All the chapters in this book investigate a specific issue related to sex trade and tourism. The introductory sections at the beginning of every chapter aims to orientate the reader and to focus his or her mind with respect to the key concepts and aspects the book covers. In addition, the book has a detailed bibliography which provides an opportunity for guided investigation where core concepts and theories are reviewed in more detail and from which the user may derive a deeper understanding of the sex tourism phenomenon.

Based on the way the text addresses the marriage between sex trade and tourism in Kenya, the book could be arbitrarily divided into two highly non-proportional parts. The first part would comprise only two chapters, seeking to establish the bases of the analysis and discussions in the remaining chapters. These two chapters present the origin and evolution of sex trade, thus providing a comprehensive introduction to the sex tourism phenomenon. The second part, of seven chapters, examines the relationship between sex trade and Kenya's tourism industry. These chapters take on an analytical flavour as they focus upon the multitude of reasons for entering the sex trade revealed in discussions with CSWs. It is obvious that the presence of tourists brings with it a variety of impacts within the destination. Therefore, the choice to pursue tourism as an agent of development needs to be made after considering all its effects across all of the planes of influence – socio-cultural, economic and environmental. The overarching objective of this part is therefore to analyse the effects of tourism on the development of sex trade in Kenya, and the coastal region in particular. Throughout the text, however, the author has resisted the hyperbole of moralists, and instead provides a disciplined framework with which to evaluate the connection between sex trade and tourism in Kenya.

The opening chapter of this book sets the scene for the more prescriptive emphasis found in subsequent chapters. It introduces the concepts of sex tourism and the need to examine it as a socio-economic undertaking. Chapter 1, 'Conceptual Background', deals with the conceptual framework of the book. The chapter opens with an overview of the importance of the development of a scientifically acceptable system of concepts to describe socio-sexual behaviours. Such concepts should have similar meanings for all analysts of the sex industry in general, and sex tourism in particular. As a consequence, tourism, sex tourism, sex tourist, CSW, sex trade and kindred concepts are explored to provide some insights into how they are used in the context of this book. The controversial issues surrounding the origins of the sex business are examined in this chapter. It ends with a presentation of the theoretical framework used to analyse the link between tourism and sex trade.

Chapter 2, 'Sex Trade and the Law: An Epitome of Legal Ambiguities', discusses the challenges of legalised as well as illegal sex trade. Reviews of the legal status of the sex trade in Kenya show that the law fails to define who fits the description of a CSW or what sex trade entails. The chapter ends with a critical look at the sex trade and the reality. Like the reminder of the text, this evaluation resists the temptation to focus on morality, and instead takes an analytical approach that matches the reality. This is crucial given that everyone seems to have an opinion on the subject of sex trade.

To lay the foundation for the study of the sex trade–tourism nexus, Chapter 3 'Tourism Development in Kenya', presents the development of the tourist industry in the country. The chapter opens with an in-depth evaluation of Kenya's tourism development during the pre- and post-colonial eras. It further discusses the process by which the government of Kenya instituted a policy of tourism promotion and expansion. It shows that the principal driving force that explains the relentless growth of post-colonial Kenya's tourism industry is the economic benefits associated with it. In consequence, the national tourism policy prioritises the large-scale economic sector. As a result, Kenya has become a major mass tourism destination in Africa. This kind of development generated undeniable levels of tourism-oriented sex trade in many destination areas (Sindiga 1995; Ndune 1996; Kibicho 2004a, 2004b, 2005b). This chapter employs Butler's (1980) model of tourism destination life cycle to locate the sex (trade)–tourism linkage in Kenya's tourism evolution process. This enables us to lay a basis for an objective examination of sex tourism in the country.

Chapter 4 'Sex Tourism in Kenya', therefore opens with a critical look at the general linkage between trading in sex and adventure travel in the olden days. This leads us to the analysis of Kenya as a sex tourism destination. The chapter reveals that sex tourists in Kenya often 'turn down' CSWs who approach them with direct sexual propositions. They prefer less explicit overtures, which give them an opportunity to continue as in non-commercial encounters, thereby creating a process which can be interpreted as confirming a mutual attraction. The chapter also examines the types of commercial sexual services offered in Kenya. It concludes with an analysis of how Kenya's CSWs advertise their (sexual) services. These techniques range from explicit to implicit forms of advertisements.

Chapter 5 'Kenya's Tourism Industry: A Facilitator of Romance and Sex' addresses the factors which have led Kenya to be among the leading sex tourism destinations in Africa. To understand the marriage between the sex trade and the tourism industry in Kenya, it is necessary to look at how tourism creates opportunities for CSWs. This chapter notes that macro-evaluations of the tourism industry usually stress the economic gains it brings to a given destination, without considering its associated socio-cultural negative effects – disruption of value systems and life patterns. The chapter provides a comprehensive overview of the motivations of sex tourists, both male and female. It then moves its focus to the influences of technological evolution, the Internet and virtual reality on the development of international sex tourism. It is undeniable that the Internet has provided an international forum where individuals can promote and sell sex tours online. The chapter thus anchors the concepts of sex tourists' destinations, surrogate tourism, cyberspace tourists and cybersex. This is followed by a detailed analysis of the role played by abject poverty in the growth of the sex trade. The chapter reports that an unparalleled poverty level is the major factor leading many Kenyans to enter the sex industry. Sex trade in Kenya, however, differs from prostitution in Western countries as there is neither a formal network of brothels nor an organised system of bar-based sex trade. Most CSWs operate independently. Nevertheless, beach boys play a pivotal role in the operations of the CSWs in Kenya's coastal region. They are the masked front of Kenya's sex tourism.

Chapter 6 'Socio-economic Base of Sex Tourism in Kenya's Coastal Region', highlights issues that influence sex trade in the region. In order to appreciate the economic significance of tourism in relation to sex trade, it is important to analyse the effects of tourism seasonality on CSWs' operations. The chapter reveals that some hotels in Kenya's coastal region are closed down during the tourism off-peak season due to low demand. Some of the laid-off workers seek temporary engagement elsewhere, including trading in sex. Consequently, this chapter investigates why people enter the sex trade profession. The reasons can be as diverse as the CSWs willing to participate in any structured sex study. However, from a general viewpoint, they range from economic to social-cultural reasons. The chapter reveals that some Kenyans make independent lifestyle choices due to the realities of socio-economic needs, while others drift into the trade through peer association and pressure.

Kenya's sex trade provides CSWs with opportunities for success and riches, while at the same time exposing them to risks to life and limb. Sex workers are vulnerable to risks such as physical attack or even social stigma. The foregoing justifies an evaluation of the sex workers' operations and how they are treated by their clients as they deliver their services. Subsequently, Chapter 7, 'Sex Trade in Kenya's Coastal Region: A Sword that Cuts with Both Edges', completes the discussion started in Chapter 6. It evaluates the challenges faced by the CSWs in the coastal region. Thus, it provides in-depth analyses of the situations in which these sex workers operate.

As the chapter's title suggests, Chapter 8, 'Sex Trade in Malindi: Roles Male Sex Workers Play', is concerned with roles male commercial sex workers (MCSWs) play in Malindi's sex industry. In its opening section, the chapter's focus of attention is on Malindi district's economy. The huge difference in economic wealth between urban and rural areas within the district is cited as the principal incentive for rural–urban migration. The second section of this chapter critically and scientifically analyses the relationship between tourism and sex trade. The findings reveal a direct perceived linkage between the tourism industry and the sex trade in the Malindi area. However, it is true to argue that male sex trade is a form of economic activity just like tourism. In any case, like tourism, sex trade in any tourism destination has a substantial local economic and social impact. Using Boissevain's (1974) distinction of entrepreneurs, this chapter reveals that the majority of MCSWs in Malindi are patrons, while the rest are brokers who make a living exclusively by manipulating second-order resources. Like most small entrepreneurs, Malindi's MCSWs grasp opportunities for profit as they spontaneously arise, gaining in every possible way from the presence of tourists without attempting to build up a stable clientele or a steadily expanding business. This is a common characteristic of small-scale entrepreneurs within the informal sectors of the national economy.

Chapter 9, 'Tourism's Influences on Sex Workers' Operations in Malindi', examines the organisation of CSWs' operations in Malindi. This case study provides a better understanding of how local CSWs try to improve both their working and living conditions through subscribing to an informal and indeed 'underground' social welfare organisation. Importantly, the author has devoted a full section of this book to the analysis of the implications of the relentless development of sex tourism in Kenya for sustainable development. The chapter therefore concludes with a detailed examination of the implications of sex trade for the development of sustainable tourism in the Kenya. This analysis is carried out bearing in mind that an understanding of future trends of (sex) tourism is imperative to enable the tourism industry to manage the future more effectively, and be prepared for the pace of change which is no longer the exception but the rule. Sustainability as an organising framework is thus an appropriate concept in responding to (sex) tourism destinations' future challenges. This section is thus instrumental in consolidating the strands and themes discussed throughout the book, and presents them in a format applicable to the development of sustainable tourism in a developing country like Kenya.

Conclusion

In closing, this introductory chapter has set the scene for a more detailed examination of the diversity of the nexus between sex trade and tourism. This is the overall theme of the remaining chapters of this book. However, it is of great importance to open this examination by first defining the key terms and concepts used in this text. This forms the principal object of Chapter 1, 'Conceptual Background'.

Chapter 1
Conceptual Background

Construction of Vocabulary for the Sex Industry

In most instances, sex has been looked at, understood, and indeed represented, as a biological urge. However, sex is more complex than just biology (Muroi and Sasaki 1997; Agrusa 2003; Rao 2003). There are socio-political underpinnings of what a society considers to be sexually ideal or deviant. Moreover, sexual desires can demand non-biological modes of fulfilment. Stimulations of various psychological faculties, for instance, can sometimes be aggressive, and in some cases destructive. Whether such stimulations are constructive or otherwise is, however, beyond the scope of the present book. Nevertheless, tourism-related sex trade or sex tourism provides a clear illustration of non-biological factors. Sex between a tourist and a local CSW, for example, 'is purely a physical encounter where the former is just but an animated object' (Rao 2003: 155). This is due to the fact that there is often minimal verbal communication, and thus human attributes like culture and family history cannot be learnt. However, CSWs use pidgin to establish their own identities as well as those of their clients.

In the attempt to create a natural science vocabulary for sex trade – that is, a system of concepts to describe typical forms of socio-sexual behaviour with identical meanings for all, or at least for all sex trade analysts – two approaches are open. First, we can invent terms, usually constructed from Latin or Greek, or both, to describe typical forms of socio-sexual behaviour. Certainly, this is the narrow way of neologism, often used by the physical and biological sciences. It is logically and scientifically the superior method, but it has the disadvantage that the larger public, and even scientists in other fields, do not understand the terms until they have been popularised. This usually corrupts their original logical purity. 'Sex trade', 'sex tourism', 'child sex tourism' and 'sex tourist' are socio-touristic examples. Second, we can attempt to put the new wine of scientific precision into the old verbal bottles of common sense. This has the twofold disadvantage that the larger public never understands the terms, and sex trade analysts themselves seldom exhibit a high degree of agreement (Cohen 1982). It thus invites endless word-wars and terminological/operational disputes. However, if sex industry analysts can agree upon the meaning of concepts as a result of significant research, this discipline will eventually prevail.

Nevertheless, there are new sex vocabularies which are becoming significant in sex industry analysis. 'Women's right to self-determination' is replacing 'women's right to freedom'; 'empowerment of women' becomes 'the equality of the sexes', while the term 'sexual rights' opens the way to 'selling sex'; 'prostitution' is replaced by 'sex trade'; 'pimps' and 'brothel-operators' are known as 'sex trade/ industry managers', while prostitutes are referred to as CSWs or simply sex workers. The CSWs' clients are also not left out of this terminological evolution, as they become 'sex service consumers' or 'sex clients'.

This chapter is devoted to beginning the debate on the relationship between sex trade and tourism development in Kenya. It introduces and defines the fundamental concepts – the building blocks – of the text, and hence reaches a clear understanding of how these terms are used in the context of this book. Lack of such clarity in usage of these concepts could obfuscate the understanding of the intricacies of the questions raised herein and what count as answers to them. In this regard, the comments by Mensah are very germane. He notes that 'clarity by itself may not be enough, but thought cannot get far without it' (Mensah 2005: 11). In striving to define some of these vague and polemical terms, however, one must guard against moralism and emotionalism. Otherwise, our analyses will be clouded by prejudice.

Definitions, Theory and Practice

Tourism

In a book dealing with sex tourism, it is sensible to begin by trying to define exactly what the concept of tourism means, before going on to examine its relationship with sex trade. However, the task of defining tourism is not as easy as it may appear, although it is relatively easy to agree on technical definitions of particular forms of tourism like safari tourism, beach tourism, conference tourism and cultural tourism, among others. The difficulty in coming up with a meaningful definition which is universally acceptable is mainly due to multi-dimensional aspects of tourism and its (in)direct interactions with other (non-)economic activities. It embraces virtually all aspects of modern society. As a result, there are many narrow operational definitions that only suit particular needs, as they are aimed at solving immediate problems (see Cuvelier 1998). Lack of uniform definitions makes it difficult to gather statistical data that researchers and governments can use to develop a database, conduct analyses and describe the tourism phenomenon.

For the purpose of this book, tourism is defined as the sum of the phenomena and relationships arising from the travel and stay of non-residents, in so far as they do not lead to permanent residence and are not connected to any earning activity. This definition thus excludes trips within the area of usual residence, and frequent and regular trips between the domicile and the workplace and other community trips of a routine character. On the same strength, the requirement that the main purpose of travel be other than exercise of an activity remunerated from the place visited is intended to exclude migration for temporary work. Consequently, a tourist is one who travels for a period of at least 24 hours in a country other than that in which he/she resides. This involves people visiting a place for sightseeing, visiting friends and relatives, and taking a vacation. They may spend their leisure time engaging in various sports, sunbathing, taking rides, photographing, bird-watching or simply sightseeing. If we stretch the scope of the tourism subject further, we may include in our definition people who visit a destination to participate in a business conference or some other professional activity such as studying or conducting some kind of scientific research. Thus, tourism involves persons travelling for health, for pleasure and/or for business. These visitors use various forms of transportation – cars, taxis, motor-coaches, campers, cruise ships, trains, motorbikes or bicycles.

In order to comprehensively define the concept of tourism, and indeed describe its scope fully, one must examine the various interested parties involved in the industry. Four global parties can be identified:

1. **Tourists** – They seek various psychological and physical experiences for personal satisfaction. These desires influence tourists' choices of destination as well as activities they participate in during the visit (the concept of tourist is defined later in this section).
2. **Tourism operators** – These are entrepreneurs who supply goods and services needed by tourists, to make a profit. They include hoteliers, travel agents, tour operators, airlines and tourist boards.
3. **Tourist host community** – In most cases, local people see the tourism industry as a source of employment. Depending on the type of tourism–local community interactions, the effects may be positive or negative, or both.
4. **Government** – This includes both the central/federal and the local/regional governments. Policymakers are interested in the tourism industry for its capacity to generate foreign exchange. Tax receipts collected from tourist expenditures, either directly or indirectly, are also of great importance for the governments concerned. In addition, many decisionmakers view the industry as a development tool due to its ability to create massive employment opportunities.

In other words, the term 'tourism' is herein used to describe four concepts: the movement of people, a sector of the economy or an industry, a broad system of interacting relationships of people and their needs to travel outside their communities, and provision of goods and services that attempt to respond to these needs by supplying (tourist) products. This description therefore views tourism as a composite of activities, services and industries that delivers a tourist experience/product.

Underlying the foregoing conceptualisation of tourism are the concepts of international and domestic tourism. Concisely, international tourism involves visits by residents of a country to another country, and visits to a country by non-residents. 'Domestic tourism' on the other hand, refers to visits by residents of a country to their own country. In this book, the term 'tourism' encompasses both international and domestic forms of tourism. Thus, *tourist* is a fundamental concept in this conceptualisation. A tourist is defined as a person on a trip between two or more countries or between two or more localities within their country of usual residence. As noted above, the purpose of the visit varies – pleasure, business, health, study, transit, and visiting friends and relatives. How long one has to spend in a destination in order to be categorised as a tourist is noted elsewhere in this section.

Tourism development brings about both positive and negative impacts on the quality of life in a destination area. On the plus side, tourism:

- creates employment opportunities;
- generates a supply of needed foreign exchange;
- develops infrastructure that helps stimulate the local economy;
- utilises local products and resources;
- diversifies the local economy;
- provokes high multiplier-effects;
- stimulates high trickle-down effects;
- promotes feelings of self-worth;
- improves the local standard of living;
- reinforces conservation of the socio-cultural heritage;
- justifies environment conservation;
- results in international understanding, and thus
- promotes a global community.

Improperly planned and developed tourism can create problems for both the host community and the environment. On the negative side of the industry, therefore, we find a number of negative effects. For example, tourism:

- causes inflation;
- enhances leakage of tourism receipts back to tourist-generating areas;
- heightens unbalanced economic development;
- generates socio-cultural problems;
- provokes excess demand for local resources;
- creates the difficulties of seasonality;
- degrades both the cultural and natural environment;
- contributes to the spread of diseases;
- increases vulnerability to economic and political changes;
- threatens family/social structure;
- commercialises culture;
- creates misunderstandings between the visitors and the visited;
- increases the incidences of crime, gambling, alcoholism, drug taking and trafficking, and sex trade.

As mentioned elsewhere in this text, any kind of development brings about change. The challenge is to achieve the right balance, which is to have tourism benefits outweigh the costs, and then come up with the best cost–benefit result. In other words, tourism and its linkages should be designed in a way to maximise net benefits for the host community. Thus, strategies and processes have to be financially feasible and socio-culturally beneficial at the same time. This said, tourism development must be part and parcel of overall economic development, which must be carried out in a sustainable manner.

Unsustainably developed tourism is referred to as *mass tourism*. This is a form of tourism which assembles a whole range of tourist activities. It is characterised by the assumption of responsibility by tourism intermediaries for a significant part of tourist activities and by the standardisation of the offers. The condition for growth in this type of tourism is the existence of mass consumption norms guided by a mass production norm, the objective being to widen access for an increasing number of individuals to cheap tourist products through rationalisation of the production costs – economy of scale. Consequently, mass tourism is practised by a great number of individuals in a specified tourist space. However, one can also qualify the presence of a small number of visitors in an ecologically fragile ecosystem as mass tourism (Medlik 1991; Cuvelier 1998; Weaver 1998; Smith and Duffy 2003; WTO 2004). It is thus a form of tourism which is offer-oriented, where the clienteles are many and anonymous. Hence, it is marked by an absence of co-production, as the tourists are relatively passive in their activities. They are satisfied by the primary tourism elements of nature, climate and landscapes available in the visited destination (Weaver 1998; WTO 2004). Price instead of quality constitutes the critical variable, and is the determining factor in this type of tourism. Competition between tourism destinations concentrates mainly on satisfying the 'standard tourist' (WTO 2004).

Kenya's tourism development, especially safari and beach tourism, from the 1980s corresponds to this type of tourism (see Chapter 3). The development of mass tourism as an economic activity, especially in the developing nations, has been associated with the expansion of sex trade (Naibavu and Schutz 1974; Agrusa 2003; Rao 2003; WTO 2004). The linkage between the two – sex trade and tourism – has given birth to a new form of tourism: *sex tourism*.

Sex Tourism

Sex tourism may be defined as tourism for which the main motivation or part of the objective of the trip is to consummate or engage in commercial sexual relations (Graburn 1983; Burton 1995; Hall 1996; Oppermann 1998; Ryan and Hall 2001). This definition, at least for this book, takes care of commercial sexual activities by both international and domestic tourists. However, this is definitely an oversimplification of the present concept, as it masks the complex process by which individuals choose to seek sexual gratification, firstly within sex trade, and secondly as a part of the tourism experience (see, for example, Burton 1995; Kibicho 2005b).

This definition, for instance, leaves a number of questions unanswered – are sex encounters with regular sexual partners while on a trip part of sex tourism? What about repeat visits to meet the same CSW in the same destination (see, for example, Chapter 9)? And sexual liaisons between fellow tourists? What of a newlywed couple on honeymoon in a foreign country? The list of these kinds of questions can undoubtedly be expanded, which definitely underscores the complexity of sex tourism as a concept (for more sex tourism-related questions, see the section 'Sex Tourists' below).

This conceptual difficulty aside, sex tourism is today a mega-industry (Naibavu and Schutz 1974; Graburn 1983; Burton 1995; Cabezas 1999; ECPAT 2002; AVERT.ORG 2004). Its covert nature often leads to insidious and uncontrollable growth, which is particularly pervasive in developing countries (Cohen 1982, 1988a, 1988b). In some South Asian countries, for instance, it generates 9–14 per cent of Gross Domestic Product (GDP) (Seabrook 1996). According to the International Labour Organisation (ILO), about US$300 million is transferred annually to rural families by CSWs in Thailand. This sum exceeds the budgets of all government-funded development programmes in the country. In the United States of America (US), a CSW can generate over US$500 in an evening (Hobson 1990). However, they usually receive less than 50 per cent of the total earnings, as a high commission is paid to the pimps.

Sex tourism can be variously voluntary or exploitative, confirming or negating a sense of integrity or self-worth (Figure 9.2) (see, for example, Ryan 1997). Non-consensual forms of tourism-driven sex trade are especially rife in many tourism destinations in Africa. They involve sexual relationships between unequal partners in terms of socio-economic status, age and gender, among other factors (Harrell-Bond 1978; Harrison 1994; Afrol 2003; Trovato 2004). Thus, some members of the host community are rendered subservient to the needs of wealthy, powerful tourists. Nevertheless, one thing remains clear: so long as tourists have money to spend, and local people have something to sell, sex included, the sex tourism phenomenon is unlikely to decline, especially in the developing nations.

Although there are myriad factors supporting the expansion of sex trade-related tourism, the Internet is a key facilitator. It has provided a convenient marketing channel for the sex industry (see, for example, Chapter 5). Websites provide potential sex tourists with pornographic accounts written by other sex tourists. These websites detail sexual exploits with CSWs and supply information on sex establishments and prices in various destinations. In addition, 'sex tour' travel agents publish brochures and guides on the Internet that cater to sex tourists. In 2004, there were over 225 businesses in the US that offered and arranged sex tours (TAT News 2004). One website promises three nights of sex with two virgin Thai girls for only US$35 (TAT News 2004: 2). The easy availability of this information on the Internet generates interest in sex tourism and facilitates sex tourists in making their travel plans. We will revisit this important subject of the role of information technology, the Internet and virtual reality in the growth of tourism-related sex trade in more detail in Chapter 5.

As noted above, sex tourism is one of the most emotive and sensationalised issues in the study of tourism. Oppermann (1998: 251) notes that the term 'sex tourism' evokes the 'image of men, often older and in less than perfect shape travelling to developing countries ... for sexual pleasures generally not available, at least not for the same prices, in their home countries'. However, in many 'exotic' tourism destinations, female sex tourism is also common (see, for example, Chapter 5). Today, sex tourism, a component of the tourism industry, has been given both overt and covert encouragement by governments as a source of foreign exchange, especially in the developing nations (Ackermann and Filter 1994; Günther 1998; Oppermann 1998; Cabezas 1999; Brown 2000). Related to the sex tourism phenomenon is the most criticised form of child exploitation – *child sex tourism*.

Child sex tourism Child sex tourism makes its profits from the exploitation of child CSWs, especially in the developing countries (Mingmong 1981; O'Grady 1992, 1994). Although such children (mainly girls) seem physically mature, they are mentally immature (O'Grady 1992). Child sex tourism also involves paedophilia-related child abuse, like filming sexual acts with minors for the production of pornographic materials (see Nick's story in Chapter 5). Further, many children are trafficked to other countries, where they are forced into the sex trade (Caplan 1984; Brown 2000). In Thailand, for example, Burmese girls as young as 13 years of age are trafficked outside the country then sold to brothel owners (AVERT.ORG 2004). While in their new 'work' situation, child CSWs live in difficult circumstances. For instance, a study by TAT News (2004) indicates that a child CSW in Thailand serves 10–30 clients per week. This leads to an estimated base of 500–1,500 clients per year.

The nature of child sex trade varies from country to country according to different gender, cultural and social factors. In Thailand, for example, 80 per cent of the child CSWs are female (TAT News 2004), while young boys account for an equivalent percentage in Sri Lanka (Rogers 1989). There are approximately 3,000 child CSWs in Kenya's coastal region, serving both international and local clients (Ndune 1996; PTLC 1998). About 90 per cent of them are girls (see Ndune 1996). Kenya's child CSWs enter into the sex industry in a variety of ways. A number of them are lured into private residences, owned by either international or local tourists, under the guise of domestic workers. In time, they are 'socialised' into sex relationships with their masters. In most cases, they are threatened or made to believe that their contracts involve satisfying the sexual needs of their employers. This practice is prevalent in Nairobi, Malindi, Mombasa and Ukunda (Kibicho 2005b). Some child CSW–tourist relationships also start on the beaches (see, for example, Joan's account in Chapter 5). This happens when unsuspecting children meet with tourists on the beach who initiate a child–tourist friendship. In other cases, tourists go through local intermediaries, especially beach boys, who lure the potential child CSW into the trap well rehearsed by the pimp in question and the concerned tourist (see, for example, Nick's story in Chapter 5).

In Kenya, the most highly organised form of child sex tourism involves employees of some well-known and, of course, politically well-connected hotels. The hotel staff facilitates the meeting between the child and the sex tourist. Such 'meetings' normally take place in the hotels, and probably with the knowledge of the hotel's management team. It is not clear, at least to this author, whether the management teams of the concerned hotels receive some percentage of the income generated therefrom. What is clear, however, is that the staff directly involved in such arrangements receive a commission from the child CSW's earnings. The nexus between poverty and child sex tourism in Kenya will be discussed at greater length in Chapter 5.

Sex tourism involves tourists from all lifestyles and all socio-economic structures. There is no single profile that can be used to describe this type of tourist. Some of them rationalise their sexual encounters with children with the argument that they are helping the children concerned and their families financially (see Kibicho 2004b). They avoid guilty consciences by convincing themselves that they are helping the implicated children to escape economic hardship. For example, Jean-Baptiste (not his real name), a tourist from Besançon, France notes:

> I am assisting these young Kenyans by paying them for their services. I have helped several of them clear hospital bills for their close relatives. Additionally, I am paying school fees for two girls. Without being cynical, is this not being mindful of the less fortunate, the needy, the poor? Anyhow, I am happy my money does not disappear into the pockets of corrupt politicians and bureaucrats. It is not being spent on senseless prestige projects. It goes straight into the pockets of the needy, the poor Kenyans. For heaven's sake, am I not helping this country?

As we shall see in Chapter 5, other sex tourists are drawn towards child sex while in the visited destination because they enjoy the anonymity that comes with being in a foreign land. This anonymity provides tourists with freedom from the moral restraints that govern behaviour in their home countries. In addition, there are myths that some people use to justify their sexual transactions with children. Some people believe that children are less likely to be HIV-positive, thus making it safer to have sex with children as it reduces the possibilities of contracting AIDS. There is even a more misplaced belief that having sex with a virgin, including infants, automatically reverses HIV-positive status to negative (ECPAT 2002; Afrol 2003; Trovato 2004). This leads to an upsurge of demand for sex with minors as HIV carriers try to cure their infection (Sindiga 1999; Afrol 2003).

In general, child sex tourism is likely to take place in and around tourism destinations where there is ignorance, endemic corruption, apathy, absence of law enforcement, and the existence of adult sex tourism. Nevertheless, tourism is not the cause of child exploitation, but the nature of the tourism setting can create an unsafe environment for vulnerable children (see, for example, Chapter 5). In other words, the existence of sex tourism in a destination might as well signal the 'presence' of child sex tourism. However, (child) sex tourism cannot prevail in the absence of the sexual services consumers – *sex tourists*.

Sex Tourists

For this book and in fact for practical purposes, a sex tourist is defined as a tourist who engages in sex with a CSW during a trip to a particular tourism destination. To put it another way, a sex tourist is an adult who travels in order to have legal sexual relations with another adult, often for the exchange of money or presents. It is the intention of this text to examine commercial sexual activities by foreign and domestic travellers together, thus this description of the concept of sex tourist encompasses both international and domestic travellers.

As with sex tourism, however, the concept of sex tourist is not contradiction-free. For example, many questions arise when one examines the concept of sex tourist: those who travel to another destination area, within the same country, for sex consumption – are they domestic sex tourists? And those interested in homosexual encounters while in the visited destination? To what extent does intention to enter into a sexual encounter while travelling need to be present to classify somebody as a sex tourist? Does it need to be pre-planned sexual behaviour? Or is openness to such a possibility sufficient? Or do both intention and the actual sex act need to be present to qualify one to be classified as sex tourist? What about tourists who visit (inter)national red light districts (RLDs) without necessarily paying for sexual services? Or is 'watching' enough for one to qualify to be called a sex tourist? In relation to this last question, it should be remembered that there are many places that offer peep-shows where clients commonly masturbate (see, for example, Chapter 4).

Moreover, travellers are classified according to their primary activity. Hence, a safari tourist is somebody whose primary goal of travelling to Kenya is to visit the Protected Areas (PAs) to watch wild animals in their natural habitats. However, the same safari tourist may also use this chance to take part in other activities – shopping, visiting friends and/or relatives, visiting other places of tourist interest, or even purchasing sexual services.

In addition, the primary motive for travel may be to experience a different physical or social environment, or to enjoy the prestige of foreign travel: 'Sex, like food, is often incidental to these things ...' (Naibavu and Schutz 1974: 66). In other words, sex activity is incidental to other travel motives (see also Oppermann 1998). Thus, if the main activity is the determining criterion, then very few tourists might be categorised as sex tourists. Similarly, if the amount of time spent on a tourism activity is the deciding factor, then there are very few sex tourists, as the time spent on sexual activities is probably not comparable to other activities.

The nature and duration of CSW–sex tourist relationships are variable (see Figure 1.2). Many of the sexual encounters with other tourists, tourism industry employees, local residents and/or sex industry workers are short-duration single events. Unsurprisingly, many sex tourists often have multiple sex partners – fellow holidaymakers, CSWs, tourist product/service providers and so on – during their travel. The prospect of this type of episodic encounter is a powerful motive for many sex tourists. However, many ongoing and long-lasting relationships have begun in the form of purely pecuniary and short-duration commercial encounters. Cohen (1982), for instance, documented many men who engaged in long-term relationships with South East Asian women. Likewise, Pruitt and LaFont (1995), noted many women from Western countries who returned to Jamaica annually to meet with their 'local lovers'. This raises the question of whether these local lovers can be categorised as CSWs.

Commercial Sex Worker

A commercial sex worker is someone who wholly or partly earns their living by providing sexual services. This definition covers all workers in the sex industry, such as erotic actors, nude models for pornography, striptease dancers, performers in peep-shows, live sex show workers, providers of erotic massage, phone sex workers and so on (see Chapter 4). In general, CSWs and their clients represent both sexes and diverse sexual orientations.

Although there are no official records on the 'global population' of CSWs, it can be estimated that the number is high. Oppermann (1998) notes that there are between 300,000 and 400,000 female CSWs in Germany serving 1.2 million customers daily; Symanski (1981) puts the number at about 250,000–350,000 in the US and more than 200,000 in Poland; Launer (1993) estimates that there are 140,000–300,000 in South Korea and 300,000–500,000 in Cambodia. According to TAT News (2004), the figure could be as high as 2,000,000 CSWs in Thailand, with up to 800,000 being under the age of 16. Kibicho (2005b) suggests a figure of 20,000–25,000 self-styled CSWs in Kenya's coastal region alone (see Figure 6.1). However, it must be emphasised that due to the illegal nature of sex trade in many countries, it is difficult to obtain accurate figures for the CSW population. Consequently, the above may be over- or underestimates.

There are many reasons to become a CSW. Many of them are related to socio-economic factors (see, for example, Chapters 6–9). In Kenya's coastal region, for example, reasons for venturing into the sex trade range from unemployment to loosening of the African traditional bonds. Out-of-wedlock childbirth and rape are other key reasons why women leave their rural areas for the anonymity of the big cities and eventually enter into trading in sex as a means to survive (Oppermann 1998). Sometimes friends, pimps or human traffickers introduce such CSWs into the sex trade (see Chapter 6). Observed critically, however, most of these reasons gravitate around (acute) poverty.

A strong consensus has established that poverty is the driving force for (tourism-related) sex trade expansion in Africa (De Kadt 1979; Backwesegha 1982; Sindiga 1995, 1999; Ndunc 1996; Afrol 2003; Jallow 2004; Kibicho 2004b; Trovato 2004). In most cases, poverty is mentioned as the principal reason for entering sex trade, both in the developed and in the developing world. According to Backwesegha (1982: 6), 'extreme poverty of Africa's rural areas and expanding urban slums is the classic springboard for sex trade'. But what is the exact meaning of the concept of poverty? A plethora of definitions exist, with each one generating more controversy than analysis of the experience of poverty. For example, there is no agreement on the number of poor people in the world, in Africa or in Kenya. In light of this difficulty, this book adopts a different approach which locates poverty in Kenya within a recognisable framework (see Figures 1.1, 1.2, 3.4, 4.1, 6.2, 9.1 and 10.1). One does not need complex econometric analysis to recognise the manifestations of poverty in Kenya. The rationale behind this methodological option is that focusing on the 'facets' of poverty allows one to bypass abstract debates and controversies surrounding the meaning of poverty, and serves as a critical point of entry into an understanding of the nexus between sex trade and tourism. It recognises that poverty in Kenya, and indeed in Africa, has evolved both diachronically (through historical time) and synchronically (within a given space in time). Further, this approach circumvents one of the shortcomings of standard measures, which fail to account for socio-cultural underpinnings that shape the face of poverty. To evade these pitfalls, this text simply defines the poor as the people living in households in the lowest income quintile, when households are ranked by five-year average total per capita income. On the positive side, this definition is based on five-year panel data, thus the income measures will be less subject to random measurement error. On the negative side, however, the definition is based on income data, not consumption expenditure data.

From a general standpoint, sex trade has always been an alternative form of work for both men and women (Finnegan 1979; Walkowitz 1980; McLeod 1982; Robinson 1989; Rogers 1989; Hobson 1990). Along with being manipulated as children, poverty is the primary economic incentive that drives people to participate in sex trade, whether directly or through following the false promises of employment offered by human traffickers. In Kenya's contemporary society, sex trade – for some CSWs – offers a good standard of income for shorter working hours, and some degree of autonomy (see Chapter 8). However, the trade is also associated with violence, criminalisation, stigmatisation, reduced civil liberties, and risk of diseases (see, for example, Chapter 7). Social stigma and criminalisation experienced by female CSWs is further compounded by the masculinist organisation and the increasing feminisation of poverty. This feminisation (of poverty) process results in part from economic, employment and social policies in most countries which have failed to fundamentally address the needs of the single female head of household. This implies that women and men, at least in Africa, experience poverty differently and unequally and become poor through different, though related, processes. Moreover, unequal sexual and social relations are ideologically and materially reciprocal, underpinned and enacted by lived relations, by jurisprudence, by socio-economic factors, and by cultural practices and processes.

Nevertheless, whether poverty is the key motivator for entering the sex business or not, commercial sexual activities in any given tourism destination are more prevalent in some areas than others. In countries where sex trade is legal, sex workers operate from officially designated areas known as RLDs. Moreover, even in countries where trading in sex is illegal, like Kenya, unofficial RLDs exist.

Red Light Districts

A red light district (RLD) is a neighbourhood where sex trade is a common part of everyday life. The term was first used in the US (Finnegan 1979). It is associated with the old practice of placing a red light in the window to direct clients to the room of a CSW, as the trade was often illegal. Earlier sex trade theorists link the term to the biblical story of Rahab, a CSW in Jericho who used a scarlet rope to help Jewish spies identify her house (Gallagher and Laquer 1987). Others claim that it originated from the red lanterns carried by rail workers during the Industrial Revolution, which were left outside brothels when the workers entered for sexual services (Walkowitz 1980).

Often the economic dimension has been the driving impetus behind permitting most forms of sex trade. However, for the desired outcomes of economic development and tourism stimulation (as in the case of Thailand and Vietnam) to occur, a large proportion of custom must come from outside the region where the RLDs are located. Alternatively, RLDs' facilities that cater primarily to locals will not have a substantial impact on local economic growth unless they draw business heavily from local residents who would otherwise leave the region in order to work.

Window prostitution is the most organised form of sex trade in RLDs. This is more so in the developed countries. In the Netherlands, for instance, it accounts for about 20 per cent of the national sex industry, and in fact it is the most visible form of sex trade in the country (Launer 1993). As the name suggests, window sex trade is a situation whereby CSWs 'display' themselves, in most cases naked, in tiny rooms visible through windows from outside. In this kind of sex trade, windows are rented for eight-hour shifts at the rate of 60–150 euros (exchange rate at the time of writing was 1 euro = US$1.281). These rates depend on the time and the location of the premises. As a general rule, rates for the 'peak' periods (weekends, and evening to midnight during weekdays) are relatively higher than those for 'off-peak' periods (morning to early evening during weekdays). Normally, the cheapest commercial sexual service in Holland's RLDs lasts 15–30 minutes. It is referred to as a 'short time', or simply a 'suck'. The price of this sexual service ranges from 40 to 50 euros. This price fluctuation is dependent on the location of the RLD. As a safety measure, all rooms are equipped with closed-circuit security systems. Most of the clients are unaware of this fact.

Some of the world-famous RLDs (sex destinations) are Hamburg's Reeperbahn, Berlin's Kurfürstendamm, the RLDs of Amsterdam and Rotterdam, Boston's Combat zone, and New Orleans's Bourbon. In Kenya, there are no 'officially' designated RLDs. As a rule of the thumb, however, all major urban centres in the country have well-known RLDs. These areas are easily distinguishable for their popularity with the CSWs, and consequently their higher levels of commercial sexual services/activities. Some of these areas are the Koinange Street, Majengo and California estates (Nairobi), the Moi Avenue/Casablanca area and Kizingo Estate (Mombasa), the Salgaa Centre and KwaRhoda Estate (Nakuru), and the Kenyatta Street, Embassy and Kisumu Ndogo estates (Malindi). Nevertheless, even within the RLDs there are some sex trade-related services which do not involve sexual liaisons between the CSW and the client. They are called *sex trade-related non-sexual services*.

Sex Trade-related Non-sexual Services

Not all clients hire female CSWs for sex. Some people hire sex workers just to have a chat. Elderly rich men who 'appreciate' the company of beautiful women commonly do this. Ryan and Kinder (1996: 511), for example, quote a (sex) tourist who observed: 'it is more important to be with a girl and have intelligent, cheerful conversation, than just to have sex'. In such a situation, however, the CSW will still be paid the same amount of money for the chat as she would if she had offered another form of sexual service. In most cases, the price is determined by the number of hours the CSW spends with the client. Sometimes the CSWs hired for chats are given huge tips. For example, Kibicho (2005b: 261) reports that 'some of them [sex workers] have been included in the wills of their elderly clients …'.

What about researchers into sex trade and CSWs' activities? They work with sex workers and sometimes hire non-sexual services. From the foregoing, therefore, it is true to argue that sex trade-related non-sexual services take many forms. However, human trafficking is the most organised kind of sexual transaction which does not necessarily involve sex liaisons with potential CSWs.

Human Trafficking in Kenya

Today, many women and children are trafficked from rural to urban areas within a given country or from one country to another (Krippendorf 1977; De Kadt 1979; Mulhall et al. 1994; Ryan and Kinder 1996; Rao 2003; Gitonga and Anyangu 2008). In most cases, the syndicate involves close relatives of the women concerned, who lure them with promises of helping them pursue education or to find them work in the cities or abroad (see, for example, Doris's case in Chapter 5).

During the late nineteenth century, efforts were made to control international traffic in women for the purpose of sex trade. Co-operation on an international scale to stamp out such traffic began in 1899 during a congress in London (Corbin 1990). This was followed by conferences in Amsterdam (1901), London (1902) and Paris (1904) (Muroi and Sasaki 1997). The principal outcome of these meetings was an international agreement providing for a specific agency in each participating country to co-operate in the suppression of international traffic in persons for the purpose of sexual exploitation (Corbin 1990; Muroi and Sasaki

1997). In 1919, the League of Nations appointed an official body to gather facts pertaining to the trafficking of CSWs. In 1921, a conference held in Geneva and attended by representatives from 34 countries established the Committee on the Traffic in Women and Children. The UN adopted the work of this committee in 1946. Three years later, a Convention on Suppression of the Traffic in Persons and Exploitation of CSWs was prepared and adopted by the UN General Assembly. It has been ratified by 72 Nations, including Kenya. The Convention defines human trafficking as:

> Recruitment of persons and/or transportation of persons by others using violence or threat of violence, abuse of authority or dominant position, deception or other forms of coercion; for the purpose of exploiting them sexually or economically; for profit or advantage of others, such as recruiters, procurers, traffickers, intermediaries, brothel owners and other employers, customers or crime syndicates. (Corbin 1990: 69)

Despite being a signatory of this convention, however, Kenya remains a principal source, transit and destination country for people being trafficked for forced labour and sexual exploitation (Gitonga and Anyangu 2008). In 2003, the country was placed on the Tier 2 Watch List due to lack of evidence of efforts to combat human trafficking (Kithaka 2004). This is mainly due to lack of law enforcement efforts, which has impeded the government's ability to deal effectively with the phenomenon. Cross-border human trafficking within the three East African countries of Kenya, Tanzania and Uganda is also on the increase (Kithaka 2004).

Trafficking of women in Kenya, especially to Europe, is well organised. There are groups who recruit young women from poor rural areas and urban centres then send them abroad. The majority of the Europe-bound recruits end up in Germany (48 per cent) and the United Kingdom (32 percent) (Sindiga 1995: 9). In most cases, recruitment agents promise the recruits lucrative jobs overseas (Sindiga 1995; Kithaka 2004; Gitonga and Anyangu 2008). The members of the syndicates make all travel arrangements for the recruits, with the understanding that they will be reimbursed once the recruits begin working. It should, however, be underlined that in almost all cases, the brokers forge travel documents for the recruits. Ironically, Kenyan bureaucrats are often in cahoots with traffickers, helping to provide false documentation for a fee.

While in Europe, the recruits are handed over to European brokers who in return pay a commission to their Kenyan counterparts. Then the European 'recruitment agents' distribute the recruits to various 'bars' and 'clubs' after payment of an agreed commission. European agents pay Kenyan recruiters three times the total costs of travel arrangements. The European bar/club owners pay European agents 12 times the amount these agents have paid to the Kenyan recruiters. Then the bar/club owners multiply the commission they pay to the agents by 1.5 when calculating the total amount of 'debt' the recruit has to pay before she or he can be 'freed'. In other words, it is the recruit who bears the costs of the entire transaction, which amounts to 54 times the total cost of the travel arrangements paid by the Kenyan recruiters. In addition to the enormous debt, the recruits are exploited within the complex human trafficking system, as there is no legal framework through which to pay the incurred debt.

In order to understand the foregoing, let us examine the visa status of the recruits. The majority of trafficked Kenyan women enter Germany or the UK through Eastern European countries (Gitonga and Anyangu 2008). They enter Eastern Europe on student visas. Like other travelling documents, these visas are in almost all cases forged (Sindiga 1995; Gitonga and Anyangu 2008). While in Eastern Europe, the European brokers arrange to transfer them to their final destinations. Accordingly, the recruits get a new set of fake travelling documents, notably 'duly' signed and stamped passports and other identification papers. At this stage, the trafficked women enter Germany or/and the UK on short-stay or tourist visas, as it is the easiest type of visa to obtain from Eastern European countries (Sindiga 1995; Bauer and McKercher 2003; Kithaka 2004; Gitonga and Anyangu 2008). It is worthwhile to note that this visa does not allow the holder to engage in income-generating activities while in the final destination country. This implies that recruits cannot look for or obtain a job legally. They work in the clubs and/or bars illegally, and thus they enjoy no legal protection. In consequence, they have to endure difficult working conditions due to the fear of deportation if caught by law enforcers. The bar/club owners take advantage of the trafficked women's insecure status to make huge profits. The recruits, for instance, receive no pay for the sexual services they offer until they finish paying off their debt. They are only provided with food and a place to stay. In some establishments, these daily expenses are added to the women's debt. Consequently, recruits take several months or even years to clear the accrued debts. Essentially, this Kenyan 'model' of human trafficking captures the general strategy employed by traffickers and brothel owners in Africa, and indeed the world over (see, for example, Muroi and Sasaki 1997; Kithaka 2004; Kibicho 2004b, 2005b; Gitonga and Anyangu 2008).

In a nutshell, a multitude of players are involved in the sex trade profession. Equally, there are diverse reasons for entering into prostitution. These facts validate an in-depth scrutiny of what the concept of sex trade entails.

Sex Trade

Skin Trade: A Historical Background

The (hi)story of sex trade is tied to the history and social construction of sexuality, cathexis and the social organisation of desire, gender relations, masculinity, and capitalist exchange relations which increasingly commodify everything, even love. However, sex trade is double-edged, and its study in contemporary society needs to address the contradictions inherent in its analysis and critique while also offering support to the CSWs (O'Grady 1994). For most CSWs, economic need is the bottom line where entry into the trade is concerned (Cohen 1982, 1988a, 1988b; Graburn 1983; Caplan 1984; Harrison 1994; Ndune 1996; Brown 2000; Jallow 2004; Kithaka 2004; Kibicho 2006c; Gitonga and Anyangu 2008).

Sex trade is often described as the oldest profession (Fanon 1966; Finnegan 1979; Walkowitz 1980). In ancient times, sex trade had religious connotations. Sexual intercourse with temple maidens, for example, was an act of worship to the temple deity (Finnegan 1979). This form of sex trade was referred to as 'sacred prostitution'. It was practised by the Sumerians. A similar type of sex trade was practised in Cyprus (Paphus) and in Corinth, where temples had several thousands of CSWs, popularly referred to as *hierodules* (Finnegan 1979). It was also common in Sardinia and in some of the Phoenician cultures, usually in honour of the goddess Ashtart. As a result of unwavering support from Phoenician leaders, the trade expanded to other ports in the Mediterranean, like Erice (Sicily), Locri Epizephiri, Croton, Rossano Vaglio and Sicca Veneria (Walkowitz 1980; Jago 2003).

In ancient Greek society, sex workers were independent and sometimes influential women who were 'officially' required to wear distinctive dresses and had to pay taxes (Walkowitz 1980; McLeod 1982). Solon, a historically famous CSW, instituted the first of Athens' brothels (*oik'iskoi*) in the sixteenth century BC (Walkowitz 1980). With the earnings from the trade, he built a temple dedicated to Aprodites Pandemo (Qedesh), patron goddess of sex trade (Finnegan 1979; Walkowitz 1980; Corbin 1990). Sex business was common in Israel too, but some prophets, notably Hosea and Ezekiel, preached against it. Consequently, there are many sex trade-related stories in the Bible. In Jericho, for example, Rahab (a CSW mentioned in the section 'Red Light Districts' above) assisted Jewish spies, and thus she was spared from death when the Israelites invaded. Salmon, son of Nahson, later married her, thereby becoming an ancestor of King David (the Holy Bible).

During the Middle Ages, sex trade was commonly practised in the urban areas (Middleton 1993). Despite the official stigmatisation of CSWs by wider society, the Church and governments have been traditionally among the institutions that benefited from sex trade (Finnegan 1979; Symanski 1981; Middleton 1993). These benefits notwithstanding, the official position of these two institutions with regards to sex trade has always been contradictory. The Catholic Church, for instance, regarded (and still regards) all forms of sexual relationships outside the marriage institution as sinful. In olden days, however, the Church tolerated sex trade because it was held to prevent the greater evils of rape and sodomy. Augustine of Hippo (354–430 AD), for instance, was of the opinion that sex trade was a necessary evil – just as a palace needed good sewers, so a city needed brothels (Finnegan 1979). In the late Middle Ages, most of the European urban governments started to control CSWs' operations within their jurisdictions. In the Languedoc region in France, for example, local governments designated certain streets as areas where sex trade could be tolerated – red light districts (Corbin 1990). As a consequence, sex trade flourished in these regions, and licensed brothels became important sources of revenue for the municipal councils (Finnegan 1979; Walkowitz 1980; Corbin 1990).

The first official acknowledgements of sex trade in Japan occurred in the fifteenth century, a period referred to as Muromachi in Japanese history, when the Shogun government started to tax 'sexual houses' within the city of Kyoto. This taxation was aimed at easing the financial difficulties the city was experiencing. In 1590, the ruler, Toyotomi, established a sex trade area in the city. He further allowed the trade in part of Osaka, in order to provide his supporters with what he termed as 'recreation and comfort' (Muroi and Sasaki 1997: 198). In the period 1603–1868, also known as the Edo Period in Japanese history, sex trade became a large-scale business in Japan. In 1617, for example:

> sex trade houses were officially opened in Old Tokyo. This was a strategy by the Government to maintain its hold on power. The war lords were strategically motivated to stay in the sexual houses. Over time, they became less wealthy relative to the growing merchant class which reduced the political tension .... Thereafter, the customers of the sex trade houses changed and private sexual houses spread throughout Japan. (Muroi and Sasaki 1997: 199)

During the colonial era, the Japanese government exported CSWs to its colonies (Muroi and Sasaki 1997). These CSWs, or *kara-yuki san*, were bonded Japanese women who were sent abroad to serve as CSWs in ports frequented by Japanese soldiers (Hall 1996; Muroi and Sasaki 1997). However, with the prohibition of legal sex trade in Japan in 1958, women from the former colonies were 'imported' into Japan as CSWs (see also Chapter 2).

In the nineteenth century, legalised sex trade became a public controversy in the UK and France (Corbin 1990). This was after the two countries passed the Contagious Diseases Acts, which made pelvic examinations for suspected CSWs mandatory (McLeod 1982; Enberg 2008). This legislation applied not only to the UK and France, but also to their overseas colonies, Kenya included (Kibicho 2004b). Many early feminists fought for their repeal, either on the grounds that sex trade should be illegal and therefore 'no government regulation is required or because it institutionalised degrading medical examinations upon women' (McLeod 1982: 4).

In more conservative times, public accommodation operators seemed to act as the last bastion of Victorian morality (Enberg 2008). Young couples checking into accommodation facilities, for example, were required to provide a marriage certificate or other evidence to show that both had the same surname (Walkowitz 1980; Bauer and McKercher 2003). All this was an attempt to deal with the growing 'sexual immoralities' which were contravening the traditional purpose of the accommodation sector. According to Bauer and McKercher (2003: 4), 'the original purpose of the accommodation sector was to provide overnight accommodation for the bona fide traveller'. The basic assumption was that these services were to be provided to singles, married couples and families. Where rooms were to be shared by members of the same sex, then separate beds for each person were provided (Bauer and McKercher 2003; Enberg 2008). But, what happened after the lights went off? Isn't what goes on in the (hotel) room only of concern to those who have paid for it? Thus, what happened behind closed doors was a guarded secret between those who paid for the hotel room. Further, entry into conventional hotels was via the reception desk, which made it difficult for clients to sneak in sexual partners.

Kenya, like other countries, has sizable domestic populations of men and women engaged in sex trade (Kibicho 2004a, 2004b, 2005b). However, until two or three decades ago, the word 'sex' had been almost like a taboo in most Kenyan communities (Sindiga 1995; Ndune 1996). It was too uncomfortable a subject to be discussed, so it was banished, resulting in an unhealthy repression and a frenetic concentration on non-sexual issues. Even today it is uncommon to have the subject discussed freely and openly in these communities. Consequently, very few, and uninformative, studies on the subject, sex trade and its different manifestations, in Kenya have been carried out. However, the dangers of this academic ignorance on the subject have been noticed. As early as 1966 Fanon, for example, warned against the poor countries becoming 'brothels of Europe' (Fanon 1966: 44). Amazingly, African and Kenyan scholars in particular have not taken such an early warning with the academic seriousness it deserved, and still deserves. In an attempt to explain this lack of academic interest in the subject, Cohen (1982) observes that any relationship involving sex trade invokes such much indignation that it neutralises any interest in serious, unbiased and systematic research (see also Chapter 4). Ironically, the scientific imperatives in the era of AIDS have only just begun to change the politics of the production of sex trade-related knowledge (Kibicho 2007).

Sex Trade Refined

Sex trade or prostitution is the sale of sexual services. As mentioned earlier in the chapter 'Setting the Scene', this definition can be expanded to capture the emotions, of the commercial sexual service provider, as well as the remuneration aspects of the sexual transaction. Consequently, sex trade can be defined in terms of emotionally neutral, indiscriminate, specifically remunerated sexual services. However, such a definition exclusively relates to sexual–monetary exchanges, where a person sells his or her body for financial gain. But what about a worker who sleeps with his/her boss for non-monetary favours? And what about a student sleeping with his/her teacher for marks – resulting in what is locally referred to as 'sexually transmitted grades' – a common trend in Kenya's public institutions of higher learning (Kithaka 2004)? Are not such persons also prostituting? These practices have become commonplace in 'modern' Kenya. The schoolgirl 'offers' herself for a few treats; the university student sleeps with his/her lecturer for a few more points; the employee becomes the boss's mistress/concubine to keep her job; the saleswoman compensates for mediocre sales by going out with some of her clients, and the wife takes other 'lovers' to make up for her husband's insufficient or non-existent income (see also Chapter 9).

Male sex workers offering their services to male customers are referred to as 'hustlers', 'rent boys' or 'punks', while those offering their services to female clients are called 'escorts' or 'gigolos' (Agrusa 2003; Bauer and McKercher 2003; Chapman 2005). In Kenya's coastal region, MCSWs (serving both men and women clients) are known as *vijana* (a Kiswahili word for 'boys') (see also Gitonga and Anyangu 2008). Gigolos escorted rich ladies of the European aristocracy to social events (Agrusa 2003). The cultural contexts through which traditional European gigolos operated differentiate them from modern Kenya's *vijana*. The traditional Western gigolo was a social phenomenon which was established by the lifestyles of the rich, thereby creating class-specific identities. On the contrary, Kenya's *vijana* operate within a tourism culture detached from a socially specific context, consequently producing classless identities.

Nevertheless, it is not the purpose of this book to distinguish between tourism-related and non-tourism-oriented sex trade in Africa. In any case, it is practically impossible to come up with a clear-cut boundary between these forms of sex trade (see, for example, Sindiga 1995; Ndune 1996; Kibicho 2006c). This is mainly because sex business in Kenya and Africa in general lies within the unregulated informal sector of the local/national economy (Chapter 5).

Depending on the cash-retention ability, sex trade in Kenya, and probably internationally, can be classified into a three-tier system as follows: (1) 'poverty sex trade', (2) 'tourism sex trade', and (3) 'high-class sex trade' (see Launer 1993; Leheny 1995; Oppermann 1998; Kibicho 2004a).

Poverty Sex Trade

Poverty turns the CSW–customer interaction into a master–servant relationship (see Figure 9.2 and Chapter 6). The CSW ventures into sex trade as a way to earn money to fulfil socio-cultural obligations, and probably as a means to eventually achieve a better lifestyle (Chapter 6). In some cases, CSWs in this category work under pimps or are attached to (that is, employed by) brothels (see Chapters 5, 8 and 9). The attachment to a second party reduces sex workers' money-retention capacity. In addition, they enjoy less 'freedom' relative to their counterparts in the other two sex trade classes (see Figure 1.1).

In Kenya, most of the CSWs operating within the bottom section of the three-tier sex trade triangle (Figure 1.1) can be categorised as street-based sex workers (see also Chapter 4). They serve all those who are interested and willing to pay for their sexual services. This group thus serves an exclusively domestic clientele. The majority of CSWs in any particular region, country or tourism destination operate within the poverty sex trade – the bottom end of the three-tier sex trade triangle (Figure 1.1) (Leheny 1995; Kibicho 2004b).

Three-tier sex trade triangle

Cash-retention capacity: Increase ↑ / Decrease ↓

Pyramid tiers (top to bottom): High-class sex trade; Tourism sex trade; Poverty sex trade

Characteristics towards the extreme ends

- Increased level of education.
- Increased physical and health security.
- Decreased number of the CSWs.
- Decreased level of competition.
- Increased level of professionalism (in terms of marketing and service delivery).
- Increased attention on client's satisfaction.
- Highly specialized market base (tourists and local elites).
- Less visible to the general public.
- Decreased number of sexual encounters per year.
- Decreased number of intermediaries between the CSW and the client.
- Decrease in the CSW's age.
- High prices per sexual service.
- Increased revenue generation.
- Poverty ceases to be a motivator.

Demarcation between high-class and poverty sex trade

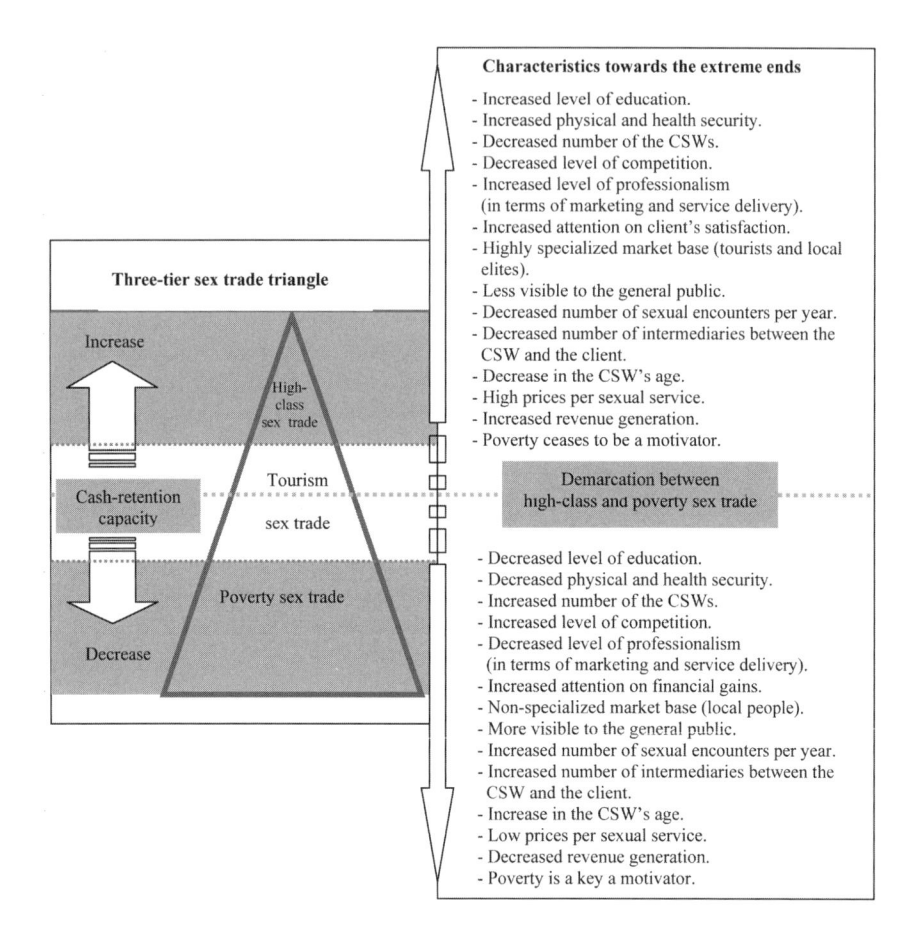

- Decreased level of education.
- Decreased physical and health security.
- Increased number of the CSWs.
- Increased level of competition.
- Decreased level of professionalism (in terms of marketing and service delivery).
- Increased attention on financial gains.
- Non-specialized market base (local people).
- More visible to the general public.
- Increased number of sexual encounters per year.
- Increased number of intermediaries between the CSW and the client.
- Increase in the CSW's age.
- Low prices per sexual service.
- Decreased revenue generation.
- Poverty is a key a motivator.

Figure 1.1   Sex trade spectrum

Tourism Sex Trade

In the tourism sex trade category, CSWs serve both international and domestic tourists. In some cases, CSWs are attached to a nightclub, a discothèque or a bar. The client has to pay a certain amount of money to the hosting club for the sex worker to be released (Chapter 4). In other instances, CSWs are independent. They carryout their sex businesses in the areas frequented by tourists, both local and foreign (see Chapters 6–9). In fact, tourists are the principal target market for sexual services in this category of sex trade.

In general, commercial sexual services are 'packaged' and 'sold' as part of the tourist product. Customers thus choose to seek sexual gratification as part of the tourist experience. Cash-retention levels for CSWs working in the tourism sex trade category are higher than those of sex workers in the poverty sex trade grouping. This 'fluid' zone separates poverty sex trade from high-class sex trade in the three-tier sex triangle (Figure 1.1). It is the sex trade 'buffer zone'.

High-class Sex Trade

In high-class sex trade, CSWs serve local elites and international tourists. The CSW–client relationship is dominated by sex–financial exchange. Sex workers' money-retaining potential is enhanced by the fact that CSWs work independently, as they are not attached to pimps or brothels. Sex workers in this category are commonly from the middle or upper classes (in their societies), who may earn a multiple of what they earn as secretaries, teachers or executives. For them, being in the sex business is more of a consumerism issue rather than the need for survival (see also Chapter 9). Overall, this categorisation system can figuratively be represented as in Figure 1.1.

It is discernible from Figure 1.1 that CSWs' level of earnings, and especially their cash-retention ability, is dependent on their position in the sex trade spectrum. The higher up one is in the spectrum, the higher the retention capacities. Towards the lower/poor end of the three-tier sex trade triangle, the number of third parties (people and institutions) taking their cut from the CSW–client transactions is large and diverse. In addition, the population of CSWs at this end is high, resulting in intense competition for clients. A combination of these factors reduces a CSW's gross earnings.

This book employs the model in Figure 1.1 to analyse tourism-related sex trade in Kenya. The model is thus heavily cited in Chapters 4, 6 and 9 (see also Figures 4.1, 6.2, 9.1 and 10.1).

Model of Transactions within a Sex Trade Community

Figure 1.2 represents a schematic model of the transactions between key actors in a sex trade community. The model considers the community of key actors in a (sex) tourism destination. Its purpose is to simplify later discussions of transactions within STCs. In the model, a STC consists of: sexual service providers (CSW), sex tourists (ST), their clients, sex trade organisations (STO), their social organisation, and tourism operators (TO), and all are embedded in society at large, which I call (E), the external environment – national/local tourism authorities, tourism trade unions, the local community, the media, the job market and so on.

The actors in the five categories in the model in Figure 1.2 interact in various ways, including interactions within groups (especially between CSWs or between sex tourists), plus all possible inter-group transactions. Interactions can be unidirectional (sex trade organisation official imposing discipline on a CSW) or a give-and-take exchange between actors (such as a CSW and a sex tourist

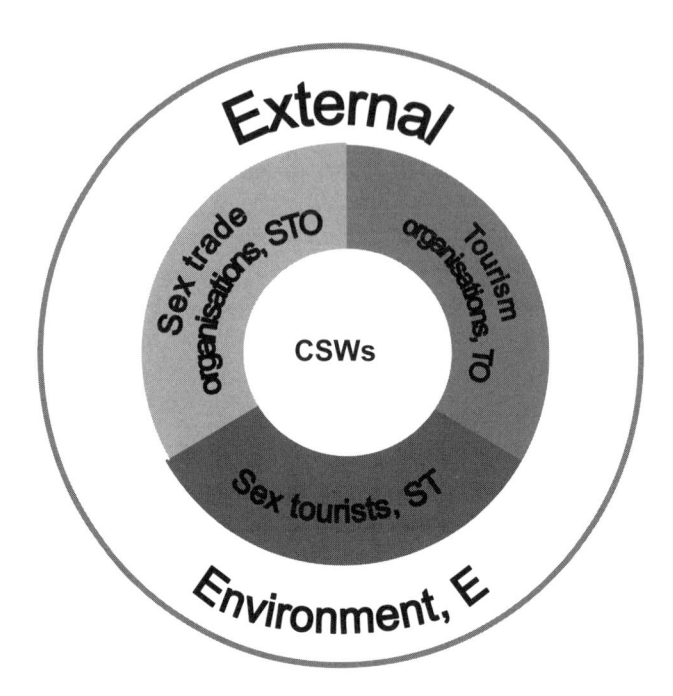

Figure 1.2 Conceptual model of transactions between members of a sex trade community

negotiating prices for sexual services to be offered). The transactions with greatest impact on the visibility of tourism-oriented sex trade are those related most closely to the sexual services delivery process: that is, CSW–ST, CSW–STO, CSW–TO, and to some extent ST–TO interactions. The social exchange theory can be used as a theoretical base for understanding the relationship between CSWs and sex tourists. The CSW–ST transaction (Figure 1.2) is linked to a history and future, emphasising a more or less durable social relationship that is modifiable over time depending on a positive/negative evaluation of the exchange consequences (Boissevain 1974; Backwesegha 1982; Burton 1995; Bauer and McKercher 2003). The three-way intersection of transactions between CSWs, sex tourists and the sex trade organisations – the CSW–ST–STO nexus – is at the heart of the linkage between tourism and sex trade in a (tourism) destination. Interactions within groups – CSW–CSW and ST–ST interactions – are also of great interest because they may constitute part of the informal mechanisms for enforcing sexual services 'contracts'. Other two-way relationships between actors will also enter the argument as relevant.

There is a general assumption about the nature of these transactions and the human beings that engage in them that needs to be made explicit. This is that human actors behave in a realistic manner. No matter how nice they may be, in the final analysis they seek their own self-interest (Fanon 1966; Jacobs 1984; Dahles and Bras 1999). According to Mensah (2005), they do so within the constraint of bounded rationality. They are not omniscient, nor can they calculate the long-term effects of all their actions (Jacobs 1984; Wels 2000; Mensah 2005). Moreover, they are capable of acting in a calculated fashion with an intention to mislead, disguise, confuse and obfuscate. Mensah (2005) calls it opportunism.

Drawing the Concepts Together

Representations of the marriage between sex trade and tourism are repeated in the print and electronic media in Kenya (Mwakisha 1995; Mungai 1998; Kithaka 2004; Gitonga and Anyangu 2008). These journalistic texts carry a common argument. Abject poverty exposes young Kenyans to sex trade; tourists sustain the demand for/and or supply of commercial sexual services by desperate locals; the Kenyan economy is to a large extent dependent upon tourism; the government is unable to address the socio-economic and political problems that have generated poverty, and thus it is incapable of controlling the expansion of (general) sex trade; for these reasons, the government is not prepared to deal with tourism-related sex trade. The circle is complete. Consequently, in the briefest terms, the conceptual framework is as follows.

First, two types of environments, internal and external, dictate the functioning of tourism-oriented sex trade. The sex tourism internal environment is what this book refers to as the STC. It comprises four general elements: CSWs, sex tourists, sex trade organisations and the tourism operators within a destination area. The external environment is the outer layer comprising all elements which have both direct and indirect influence on the internal environment's/sex tourism's operations. They include, but are not limited to, the global market, government policies, the tourist host community and societal trends. Thus, external actors include state or national/regional/local tourism authorities, tourism trade unions, the local community, media (both print and electronic), employers and the job market, and such factors as tourism policies, sex trade laws and general poverty levels.

Second, faced with unceasing economic uncertainties, Kenya's post-colonial government embraced tourism as a means to diversify its economy and ameliorate poverty levels among its people (see Chapter 3). As a result, all levels of Kenya's government involvement in tourism typically focused on its economic aspects. Thus, priorities for the types and the magnitude of public sector financial support of national tourism developments were established through a range of government policies. These policies favoured foreign-dominated, large-scale (mass) tourism development.

Third, tourism is an agent of social change in tourist host communities. Whether it initiates it or is simply a facet of ongoing change is a subject of debate.

However, one thing is certain, as the industry evolves through exploration to involvement, involvement to development and then development to consolidation stages or from small-scale locally owned to foreign-oriented mass tourism, it brings both (non-)economic benefits and costs to host communities (see Butler 1980; Dieke 1991; Honey 1999; Kibicho 2003). Nevertheless, many locals seek to profit (economically) from the flourishing tourism industry. Owing to unparalleled levels of unemployment in Kenya, however, tourism cannot absorb all the local jobseekers willing to work in various capacities in the industry. Faced with these hard realities, many of them design other means to benefit from the industry, including trading in sex (Naibavu and Schutz 1974; Cohen 1982, 1989; Launer 1993; Leheny 1995; Ndune 1996; Cuvelier 1998; Dahles 1999; Brown 2000).

Fourth, it is apparent that physical security and comfort are important motives, but not the only ones for human action. The goals individuals seek are determined by their needs and their values. In the hierarchy of needs, physical survival, security and comfort are primary (Maslow 1954). Vital social and mental needs gain prominence when the basic physical needs are met. As an individual prospers, the vital urge for intensity, excitement, enjoyment, adventure and self-expression becomes more important. Beyond these lies the mental urge for curiosity, knowledge, creativity and imagination. Unfortunately, despite their enormous reservoir of human energy, most of the CSWs in Kenya are struggling at the base of Maslow's pyramid of needs (see also Figure 1.1). Abject poverty and all its concomitants drive young Kenyans, both men and women, into tourism destination areas to seek work. Faced with unending social pressure to support their needy families and dwindling prospects of securing formal jobs in the tourism industry, many Kenyans venture into trading in sex as their last option (see also Finnegan 1979; Findley and Williams 1991; Kibicho 2004a, 2004b, 2005b, 2006c). In such cases, it would be illogical to view tourism as a sole creator of sex trade. Anyhow, the bottom line is that any industry or economic activity that condones exploitative labour conditions and income instability cannot be considered pro-poverty alleviation. For these reasons, Africa's tourism development strategies cannot be focused solely on growth. They need to take into account likely distributional implications and be grounded in a solid understanding of who the poor are and why so many of them have difficulties in breaking out of the vicious circle of poverty.

Viewed differently, tourism can be a tool for poverty alleviation. However, it is worth noting that a tool may be used to perform or facilitate a task, but it cannot compensate for ill-conceived plans, lack of capacity and/or co-operation and general dysfunction (see, for example, Jacobs 1984). It cannot solve problems associated with corruption, nor can it rid a destination of cronyism (see Chapter 3). This said, the central thesis of this theoretical framework is that it would be foolhardy to think that there is one simple solution. To truly overcome the ill effects of poverty sex trade (Figure 1.1), government efforts need to reduce the number of CSWs driven into sex business by poverty, not create ways to cope with results.

First, leading on from the foregoing, regular police crackdowns backed by sex trade legislation which is situated within a public nuisance framework only address the symptoms, leaving the socio-economic situations generating the trade intact. From this perspective, the aim is reduction of public nuisance. Consequently, this approach concentrates excessively on the narrow issue of how to deal with the CSWs, as they are seen as the cause of sex trade (Sindiga 1995; Ndune 1996; Kibicho 2004a, 2005b). Thus, the laws criminalise not the trading in sex, but rather the activities that surround it – soliciting, kerb crawling, procuring, brothel keeping and living off the earnings of a CSW (see Chapter 2). This renders the application of the sex trade laws confusing and ineffective. This ineffectiveness is exemplified by the routine shuttling in and out of court and jail of hundreds of Kenya's CSWs on weekly bases, without any significant impact on the general manifestation of sex trade. Nevertheless, a couple of countries have designed realistic and coherent sex trade laws (see Chapter 2). Policymakers in many countries, for instance, have watched the implementation and impacts of New Zealand's new legislation with great interest (Rao 2003; Jallow 2004). This law is based on the perspective of sex trade as work. Consequently, it is more realistic and aims at enhancing rather than curtailing human potentials (Naibavu and Schutz 1974; Delacoste and Alexander 1988; Cabezas 1999).

Second, the high numbers of CSWs willing to quit sex trade in favour of alternative economic undertakings is a clear revelation of the huge reservoir of potential human energy existing within Kenya's leading tourism destination areas. If government policies, strategies and programmes can tap this latent energy and channel it into constructive activities, they can stir the regions concerned to action and rapid advancement. Such initiatives could lead to a qualitative enhancement in the capabilities of the STC which will help to break the existing patterns of social behaviours and form new ones. However, it is worth noting that the growth and evolution of tourism-related sex trade is complex in nature, so intervention programmes which seek to effect change systematically need to be based on an organised understanding of its dynamic and complex processes.

Third, prevailing attitudes of community members are among the most frequent and powerful influences on whether people adhere to informally established norms and standards of behaviour. In this sense, the local community is a 'principal', and the actors within its tourism destination areas are its 'agents'.[1] This is clear with regard to community political decision on how to deal with sex trade in general. The community also includes the media that influence attitudes towards sex business and local tourism authorities that set requirements and standards of which influence what happens in a tourism destination. The degree to which the community sets its priorities with regard to trading in sex and the value it places on tourism development has a direct influence on the STC's operations. In any

---

1 Principals are persons/establishments of authority who want something done, while agents are those who perform the desired activities (Boissevain 1974; Jacobs 1984; Dahles 1999).

case, a community is a subconscious living organism which strives to survive and develop (Boissevain 1974; Jacobs 1984; Mensah 2005). Individual members express conscious intentions in their words and acts, but these are only surface expressions of deeper subconscious drives that move the community at large.

Fourth, an idea that lies at the core of this book is that sex tourism is an economic activity. Thus, CSWs and the energy and effort they exert in the process play a central role in its wellbeing. Sex workers and sex tourists are therefore involved in a process of 'co-production', in which both the producers – CSWs and sex trade organisations – and the consumers – sex tourists – must work together if they are to realise their individual, and sometimes mutual, goals. Viewing CSW–sex tourist interaction, sex tourism, as a process enables us to address questions somewhat differently than the more traditional view of looking at CSWs as solitary producers (see, for example, Walkowitz 1980; Truong 1983; Smart 1992; Sindiga 1995; Ndune 1996; Ryan and Hall 2001). Consequently, this study incorporates the notion of systems[2] view in its conceptual framework in order to model the dynamic and complex nature of the interactions which occur between the multiple actors within a STC. This view will ensure that any policy targeting sex tourism adopts a framework which integrates all knowledge of factors, instruments, conditions, agencies and processes of the industry's development. Rather than singling out a specific set of determinants or giving primacy to a limited set of instruments/conditions/agencies, the approach would reveal the nature of the relationships and motivations that govern the interactions of the key actors (see Figures 1.1 and 1.2). Thus, it will be possible to view the whole sex trade phenomenon from multiple perspectives that are integrated and unified, rather than considering individual 'parts' of the STC. Moreover, human beings are gregarious, so as time goes by, groups (sex trade organisations) evolve through several stages, with CSWs developing more complex bonds and relationships (see, for example, Chapter 9). Over time, evolving group norms shape individuals' identities and behaviours. Leading on from the above, therefore, the system approach is important, as sex tourism is a complex, multi-dimensional whole that evolves in many inter-related directions simultaneously.

The fundamental hypothesis underlying the model presented here is that any genuine government has the capacity to direct the flow of social energies through the instrumentation of law, administrative procedures, controls, incentives and public policies. However, scepticism and caution towards such a complex conceptual model is likely warranted, particularly for a field such as (sex) tourism which has thrived on simple principles.

---

2   A system is a set of inter-related sub-systems co-ordinated to form a unified and organised whole to accomplish a set of goals.

Conclusion

This chapter has set forth the conceptual framework within which to examine the relationship between tourism and sex trade in Kenya. This theoretical framework, including the concepts of tourism, sex tourism, sex tourist and sex trade, provides the basis on which to examine this linkage. The next chapter seeks to start our journey of bringing the theory into contact with reality. It starts with an in-depth evaluation of the legal status of sex trade.

Chapter 2
Sex Trade and the Law: An Epitome of Legal Ambiguities

Introduction

As long as large numbers of people live and work together in close proximity, there will always need to be a system of rules or laws – formal or informal. The purpose of laws is to provide consistency of application and equality of treatment, which ensures that fairness and justice exist (Jackson 1986). In other words, laws are established for the benefit of the common good. The mechanisms for enforcing rules/laws are usually formal. They include the network of legal provisions that establish and guarantee that laws will be observed – the court system ensures that laws are obeyed, and the law enforcers ensure that laws are respected. If the enforcement works, it can be relied upon to see that the laws are observed (Jackson 1986; Cabezas 1999). This reduces uncertainty and provides a degree of confidence (see Chapter 9). If rules break down and there is poor enforcement, things do not work as intended, and there is a feeling of insecurity and doubt about what one should (not) do. Where laws are weak or contradictory – where socio-economic activities tend to be chaotic or there is little respect for fellow community members, including CSWs – then the environment for tourism development is damaged (McLeod 1982; Plant 1990; Ackermann and Filter 1994; Pruitt and LaFont 1995; Kempadoo 1999). In extreme cases, members of a community may be anxious about their safety. Consequently, formal laws have far-reaching effects on sex trade, despite being categorised as an illegal business (Leheny 1995; Phillip and Dann 1998; Kibicho 2004a, 2004b, 2005b).

The foregoing explains why we have devoted a whole chapter on the discussion of sex trade and law. In so doing, the book adopts an international outlook when examining sex trade laws. This perspective enables us to exhaustively examine the legal implications of tourism-oriented sex trade, which is today an international phenomenon (Fanon 1966; Harrell-Bond 1978; Mathews 1978; Cohen 1982, 1988a; Graburn 1983; Pruitt and LaFont 1995; Cabezas 1999; Afrol 2003). Moreover, sex trade laws in Africa have been shaped primarily through different inheritance of laws which existed in Europe through the eighteenth century (Jackson 1986). Thus, Kenya's sex trade law has a mixed legal system, made up of the interweaving of a number of distinct legal traditions – the common law system from its English coloniser, and African customary laws. However, these traditions have a complex inter-relationship, with unending influence by laws from other countries and institutions, notably the US and the United Nations. Consequently, an analysis of a boundariless socio-economic phenomenon like tourism-related sex trade can only be successful if it takes these inter-relationships into account.

Taking an international perspective, therefore, the main goal of this chapter is to examine legal contradictions concerning the regulation of sex trade. It looks at how legal ambiguities have enhanced the growth of sex trade in Kenya, while at the same time exposing CSWs to difficult working conditions (see Chapter 7). These difficulties are due to the fact that Kenya's laws, as in the rest of Africa, criminalise only the acts associated with selling sex, and not the trade itself (see Chapter 1). Thus, sex workers are subjected to various forms of 'punishments', while their clients get off scot-free (see, for example, Chapter 7). In many occasions, for instance, female CSWs in Kenya are punished with fines and/or prison sentences, but not their pimps and procurers (Backwesegha 1982; Sindiga 1995, 1999; Ndune 1996; Kibicho 2004a). This chapter therefore sets the pace for the analyses carried out in the subsequent chapters.

Attempts to Control Sex Trade around the World

As a result of the epidemic of STDs in Europe in the sixteenth century, different strategies were designed with the aim of controlling sex trade (Finnegan 1979; Walkowitz 1980; Stanley 1990; Hobson 1990; Kempadoo 1999). Thus, brothels were closed while at the same time CSWs were subjected to severe punishments (Walkowitz 1980). When these measures proved unsuccessful in eliminating sex business, some European cities developed more stringent measures against CSWs and sex trade in general (see also Chapter 1). Berlin, for instance, required medical inspection in 1700, while Paris began to register its CSWs in 1785 (Walkowitz 1980). In the UK, legislation to control the spread of STDs was embodied in a series of Contagious Diseases Prevention Acts (1864, 1866 and 1869) requiring periodic medical examination of all CSWs in military and naval districts (McLeod 1980; Walkowitz 1980; McLintock 1992). It also called for detention of the CSWs infected with STDs (Walkowitz 1980). These Acts were repealed in 1886 because they had failed to control the spread of STDs (Stanley 1990). In 1898, the Vagrancy Act prohibited males from living on the earnings of female CSWs (see, for example, Walkowitz 1980; Stanley 1990; Symanski 1981). It is worth reiterating that like in other British colonies, Kenya's sex trade was governed by these Acts (Jackson 1986).

By the mid-nineteenth century, sex trade was widespread in the US, too (Finnegan 1979). However, the trade was thought to be connected with crime and general insecurity (Hobson 1990). Nevertheless, from a legal point of view, the federal government seemed unbothered by its existence. For example, there was no 'legal' effort to deal with the trade until towards the end of the nineteenth century. In 1910, the Mann Act, popularly known as the White Slave Traffic Act, was enacted (Finnegan 1979; Hobson 1990). The Act forbids inter-state and international transportation of women for immoral purposes. Like many laws on sex trade, this Act does not say what the term 'immoral purposes' means. In consequence, its interpretation depends on who is being judged and who is judging (Hobson 1990; Jago 2003; Kempadoo 1999). By 1915, nearly all the US states had passed laws regarding the operations of brothels and other aspects of the sex profession. None the less, during the First World War there was an increase in sex trade near US bases which led the American Congress to enact the May Act in 1941 (Finnegan 1979). This law made it a federal offence to offer commercial sexual services in areas designated by the secretaries of the US army and navy (Mathews 1978; Finnegan 1979; Hobson 1990).

Legal Quandary

All US states except Nevada have legislation that makes it a crime to operate a brothel (McLintock 1992; Kempadoo 1999). Further, most states have laws against all forms of sex trade, although they often exempt CSWs' clients from prosecution. To illustrate the legal ambiguities inherent in the US sex trade laws, let us consider the *Latex nipples case* of 1994. In late 1993, a sting operation was undertaken in Louisiana by law enforcers (Jago 2003). During this operation, some striptease dancers were apprehended. According to the charge sheet, the crime they had committed was 'dancing with their nipples exposed' (Jago 2003: 129). Louisiana's obscenity statute obliges women, including striptease dancers, to always cover up their nipples. However, the law does not detail how and to what extent this covering should be done (Jago 2003). Consequently, the strippers capitalised on this lack of clarity to defend themselves. They argued that:

> they had masked their nipples in compliance with the obscenity law by painting on a layer of latex, allowing it to dry, then applying a foundation powder, and an appropriate cosmetic make-up to mimic a nipple. (Jago 2003: 130)

They claimed that their nipples were not displayed to the spectators without the presence of a substantial cover.

To enable members of the jury make an informed decision on whether the accused had 'appropriately' covered their nipples, the accused were required to apply the described make-up in the anteroom, and then return to the courtroom. Some minutes later:

> they returned and as per the Jury's instructions, they bared their breasts. At the end, the Jury found it difficult to determine, especially considering the lighting of a typical nightclub, if the 'offending nipples' were uncovered on the relevant night or if, as the dancers claimed, they were covered by the demonstrated replicas. (Jago 2003: 130)

The strippers were found not guilty of all the charges. Similar lack of clarity also exists in many European countries' legal systems.

To demonstrate these ambiguities, let us consider the case of Germany, where sex trade was legalised in 2000 (Chapman 2005). According to Germany's sex trade laws, brothel owners are supposed to pay government taxes as well as employees' health insurance. Subsequently, brothel operators are granted access to the government's database on jobseekers. Further, under Germany's welfare reforms, any woman under 55 years of age who has been unemployed for more than a year must accept any available job, including jobs in the sex industry, or lose her unemployment benefit (Chapman 2005). The government had considered making brothels an exception on 'moral grounds'. However, the authority found it difficult to distinguish them from bars. As a result, job centres are required to 'treat employers looking for CSWs in the same way as those looking for dental surgeons' (Chapman 2005: 11). The new regulations state that 'working in the sex industry is not immoral, and thus those who refuse to take jobs in the industry will lose their unemployment and other related social benefits' (Chapman 2005: 12).

In 2004, for example, a 25-year-old unemployed information technology graduate lost her unemployment benefits for not taking a job in a brothel. She had registered herself at a job centre as a jobseeker. According to the job centre's records, she had previously worked as a waiter in a café and she was willing to work as a night barmaid. A week later, she received a letter from the centre telling her that an employer was interested in her profile. After calling the potential employer to arrange an appointment for a job interview, she realised that it was a brothel which was interested in her services. She was accordingly supposed to offer sexual services to the brothel's customers. Consequently, she turned down the job offer. Because of this decision, she lost her unemployment benefits (Chapman 2005).

When she started the process of suing the job centre concerned, she found out that the centre was operating within the law. Job centres that fail to penalise people who turn down job offers by instigating the process of stopping their social welfare benefits face legal action from the potential employer. Conversely, advertisers can sue job centres that refuse to advertise sex trade-related jobs (Chapman 2005). This German scenario is a textbook example of how sex trade laws can sometimes be insensitive to some members of society.

Sex Trade's Legal Status: An Acrimonious Debate

It is an undeniable fact that sex trade takes place just about everywhere (Harrell-Bond 1978; Thompson and Harred 1992; Cabezas 1999; Ryan and Hall 2001; Jallow 2004). It is not restricted to certain societies, areas or seedy parts of a city. It is only more apparent in some places than in others, either because it is in fact more prevalent, or because it is in a form that attracts more public and police attraction (Symanski 1981; Kempadoo 1999; Kibicho 2004a, 2004b, 2005b; Trovato 2004). In response to this fact, many countries, Kenya included, have designed laws against sex trade. However, CSWs have learnt how to carry out their 'illegal' business without or with minimal confrontations with law enforcers (see

Chapters 7–9). Thus, there is a consistent growth in the trade, especially in many urban centres and tourism destinations. Despite well-stipulated laws, in Thailand, for example there are conservatively estimated to be 300,000 CSWs in Bangkok alone (TAT News 2004). The country has a reputation for being the sex capital of Asia, with sex trade-oriented tourism reaching its peak during the 1970s and early 1980s (Cohen 1993; Launer 1993; Leheny 1995; Phillip and Dann 1998), but declining since the threat of HIV-AIDS began to change visitors' attitudes (Cohen 1988b). In Kenya, a similar situation prevails, although there has never been an official estimate of CSWs (see Chapters 4–6). Consequently, unofficial estimates vary considerably according to the motives of the authors, whose works are rely heavily on secondary data (Kibicho 2004b, 2005b).

Anyhow, the sex industry has taken on an international dimension (Cohen 1982; Sarpong 2002). It has been recognised as an economic motor for a number of countries, particularly in Asia (see Chapter 1). The irony is that even in those countries where the sex industry plays a significant role in national economic wellbeing, trading in sex is not entirely legal. The critical questions at this stage are: Would legalisation of the sex trade reduce some of the inequalities and abuse suffered by CSWs? Would legitimisation of the trade negate the achievements realised in the last couple of decades in the spheres of promotion of human rights and improvement of women's, and girls', status? Therefore, the issue here should no longer be a question of morality, but rather a practical (re-)examination of the reality. It is only through a rational and non-moralistic approach that long-term and socio-economically acceptable alternatives can be found (Kempadoo 1999; Bauer and McKercher 2003; Enberg 2008). Is sex trade a vice, and are those involved in it evil or somehow lacking in judgement? Are CSWs passive victims who should be guided and saved? This leads us to the double-edged question: Is sex trade a form of exploitation to be abolished, or an occupation to be regulated? On the surface, this looks like rehashing a timeless debate.

The question of whether sex business is a form of exploitation or an occupation is the most emotive, and indeed divisive, issue among women's lobby groups throughout the world (McLeod 1982; Caplan 1984; Cabezas 1999). There are two camps – 'abolitionists' and 'regulationists'. On one hand, the abolitionists advocate for a total ban on the sex trade as it degrades women (Stanley 1990; Maurer 1991; Enberg 2008). They argue that the regulationists hypocritically represent the interests of the pimps, the human traffickers and the general procurers, with no regard for CSWs' welfare (Stanley 1990; Findley and Williams 1991). On the other hand, those who are in favour of regulation of the profession view people in the sex business as workers with something to offer humanity, who thus should be treated with dignity (see also Chapter 6) (Symanski 1981; McLeod 1982; Jago 2003; Chapman 2005). To them, CSWs are creative and social beings who make class contributions to their societies and to general civilisation (see, for example, McLeod 1982; Cabezas 1999). They feel that the abolitionists are locked away in the ivory towers of academic feminism, where they are totally cut off from the day-to-day realities facing CSWs (see, for example, Symanski 1981; McLeod 1982).

The dividing line between the abolitionists and the regulationists lies in the distinction between 'free' and 'forced' prostitution (Jago 2003; Enberg 2008). Abolitionists generally maintain that the vast majority of sex workers are forced into sex trade, while their opponents insist that this is not necessarily the case. Leslie (not her real name), a CSW in Mombasa (Figure 6.1) who seems to subscribe to the anti-abolitionist group, categorically warns:

> Assuming that all CSWs are forced into sex trade is very passionate – a position from which a country cannot make its policies. Public policy has to be founded on a good analysis; it should not be faith-based. Before trying to clean up [remove sex workers from] the streets and beaches, one needs to look into the causes of the sex trade and its structures. There are definitely complex socio-economic factors which have lead us [CSWs] into this trade. Moreover, there are hidden sex trade-related networks and activities which cannot be dismantled by the uncalled for harassment of hard-working Kenyans, through police crackdowns done in a callous and disrespectful manner. We [CSWs] are exhausted from the constant threat to our livelihood.

> If we [CSWs] are barred from working from the streets, procurers will use some of us [CSWs] in hidden/secret premises, and in fact in places which are dangerous for those involved. At the end of the day, the government will be creating more problems for us [CSWs], our clients and our dependants ... and in any case the number of CSWs is growing in tandem with the increasing number of customers. Sex business is demand-driven. Why then is the effort to combat the trade one-sided? The government should also target the demand side of the business. Or does it mean you can have supply if you do not have demand? This is certainly not rocket science, it is simple economics.

Fatima (pseudonym), another CSW in Kilifi (Figure 6.1), completes Leslie's story:

> I was working as a sous-chef in a tourist hotel. However, I was not making enough money to meet my daily needs, so I needed to augment my income with a part-time job. I became a striptease dancer. I would be up and down those poles and up and down on my clients' laps all night, and often until five o'clock in the morning. It became awfully tiring for me since I had to be in the hotel from 8H00 to 17H00 (local time). I realised that I had to stop moonlighting and choose one of the two occupations. I chose exotic dancing, for it pays more with less stress.[1]

From these pertinent issues raised by Leslie and Fatima, it is clear that the free/forced distinction of the sex business oversimplifies the socio-economic and

---

1 See also Zainabu's comments in Chapter 4.

political aspects of sex trade in Kenya, and in Africa in general. To understand the over-simplification inherent in the free/forced differentiation of sex trade, consider the following three additional scenarios. A 14-year-old girl attached to an AIDS-infested 'nightclub' (read brothel) in Salgaa Centre (read RLD) in Nakuru clearly never consented to this form of sex exploitation (Gitonga and Anyangu 2008). A drug addict at Kizingo Estate (read RLD) in Mombasa who must serve a predetermined quota of clients to get her daily share of drugs from her pimp is not free to make decisions concerning her body (Kibicho 2004a). What of a Ugandan woman who loses her job in her home country and decides to go to Malindi to work as a barmaid, but ends up working as a street CSW (Kibicho 2005b)? Can she make an informed decision on what or what not to do with her body?

Certainly, for these CSWs in Kenya, the distinctions between free and forced sex trade obscure the powerful structural socio-economic conditions that often drive Kenyans into the sex industry. Among others, such conditions include poverty, marginalisation, lack of employment opportunities, and prior sexual abuse (see, for example, Chapter 6). In order to understand the foregoing, we need to address the following three fundamental questions in an objective manner: When does anyone make free decisions, especially in the modern capitalistic labour market? A man who works in a coffee factory, whose wages will never get him over the poverty line – did he choose this way of life? What about a woman whose background never afforded her the chance to develop any skills? Can she really discriminate between the glamour of money and the promise for a good life when she has no other means to make ends meet? Coupled with economic crises, natural disasters, political unrest and ethnic conflicts, these situations make many Africans vulnerable and easy targets for recruiters and human traffickers (see Chapter 1).

However, irrespective of the sex trade camp one subscribes to, abolitionist or regulationist, a critical examination of CSWs' reality in Kenya and elsewhere in the world reveals that a total eradication of the sex trade is more challenging than theoretically documented (Symnaski 1981; McLeod 1982; Maurer 1991; Ndune 1996; Oppermann 1998; Kempadoo 1999; Sindiga 1999; Kibicho 2005b). To show the challenges inherent in the abolitionists' approach, let us briefly consider the development of sex trade in Korea and in the South East Asian countries in the late 1970s. Sex workers in Korea were required to obtain professional identification cards from the government in order to enter hotels (Muroi and Sasaki 1997; Jago 2003). While undertaking the government-designed orientation programme for CSWs, which was a prerequisite for the issuance of the identification cards, women were 'taught' that their carnal exchanges with foreign tourists did not prostitute themselves or the nation, but were rather an expression of their heroic patriotism (see, for example, Rogers 1989; Hall 1996). Incensed by these teachings, the Christian Women's Federation of Korea formulated and publicised a declaration opposing all forms of sex tours in the late 1970s (Muroi and Sasaki 1997; Jago 2003). This culminated in a series of nationwide demonstrations. Consequently, male Japanese sex tourists shifted their base from Korea to South East Asian countries, where sex tourism flourished in the early 1980s (Rogers 1989; Muroi and Sasaki 1997; Jago 2003). However, the protest campaigns eventually spread to the South East Asian region (Muroi and Sasaki 1997). The protestors' activities were aimed at a total ban of sex trade, presuming that the trade was offensive and exploitative towards CSWs. It was a condemnation not of the individual sex workers, but rather the sex profession as an institution.

These demonstrations put effective pressure on the Japanese Government. Accordingly, the government, through the ministry in charge of tourism, withdrew operating licences from tour companies and the travel agents who had organised the 'offending' sex trips (Jago 2003). Overall, the criticisms and protests faced by male Japanese tourists decreased the numbers of sex tours to South East Asia (Phongpaichit 1981; Cohen 1982, 1988b, 1989; Robinson 1989). As a consequence, South East Asian women migrated to Japan, where they worked as CSWs of all kinds – striptease, call girls, street sex workers and brothel-based CSWs (Cohen 1982, 1988a; Findley and Williams 1991; Muroi and Sasaki 1997; Jago 2003).

The point here is that when male Japanese sex tourists were unwelcome in Korea, they shifted to South East Asia. Further, when the sex trade environment became unfavourable in this new sex destination, female CSWs followed sexual service consumers to their home country – Japan. In other words, CSWs migrated to search for jobs, notably in the sex industry (Findley and Williams 1991; Bauer and McKercher 2003). In this case, CSWs became the demand and the supply at the same time.

Moreover, although women are often thought of as 'passive movers', migrating only to join or follow family members, research has found that economic rather than personal considerations predominate (Muroi and Sasaki 1997). In most cases, young women migrate in order to help their families by sending money home more often than men, although their earnings are usually lower. Without justifying the movement by female CSWs to Japan discussed above, it should be noted that apart from the foreign currency which is remitted to the migrant workers' countries, migration has two additional advantages:

1. It increases employment opportunities for the people who remain in the country.
2. The ex-migrant workers transfer technologies back to their home countries. (Findley and Williams 1991)

The question for the abolitionist camp is this: By abolishing sex trade in a given area without offering those involved alternative economic undertakings, are we not encouraging the CSWs concerned to shift their operation base(s) elsewhere – just like the Korean's CSWs?

Regulated Sex Trade

At one end of the legal spectrum, sex trade carries the death penalty in countries like Saudi Arabia and Iran, while at the other end CSWs are tax-paying and unionised professionals in countries like Germany, the Netherlands and New Zealand (Truong 1983; Jago 2003; TAT News 2004; Chapman 2005). In most of the countries where sex trade is legal, operation of brothels as well as advertising sex businesses are legal. However, CSWs must be at least 18 years of age, while for non-commercial sex, the legal age of consent is 16 years (Pickering and Wilkins 1993). These regulatory approaches are undertaken with the recognition that sex trade is difficult to eliminate (Chapman 2005). In consequence, some countries have chosen to regulate it in ways that minimise undesirable consequences. The goals of such regulations include controlling the spread of STDs, reducing forced sex trade, and weakening the link between sex business and criminality.

Essentially, legalisation of sex trade invokes changes in the way the sex industry is organised, and ultimately how it operates. For instance, licenses are issued with mandatory health check-ups. Laws then set up 'erotic zones' or RLDs away from residential areas, while at the same time spelling out the standards with regard to CSWs' working areas. Such standards include sufficient lighting, quality mattresses, fire precautions and air supply (see Chapter 1). In addition, appropriate tax brackets are determined, rights to health insurance ensured, and retirement benefits guaranteed. These regulatory measures are intended to protect CSWs as well as their clients (Chapman 2005).

However, as we have seen in the cases of the US and Germany, even in the countries where sex trade is legal, there exist some legal ambiguities, and at times the law is anachronistic. Rules, for example, vary as to which sex trade-related activities are illegal – being a CSW, being a client, being a pimp, advertising sexual services, soliciting (offering sexual services for money), owning or operating or even working in a brothel. Sweden, for instance, outlaws buying, but not selling of sex (Corbin, 1990). In the case of the Netherlands, it is legal for somebody under 18 years of age to be a CSW provided his or her client is not under-age, below 16 years (Corbin 1990; Cabezas 1999). In Canada, sex trade is legal, while soliciting, in public, is not. One can live on the earnings of one's own sexual services, but may not live on those of another CSW (Baillie 1980). It follows logically that if one cannot live on the income of a CSW, the children supported by mothers who earn a living by trading in sex, and the husbands or boyfriends of such women who share food, clothing and shelter paid for through sex trade-related earnings, are also guilty of a criminal offence – although they play no active role in the act of sex trade itself. Due to its refreshing directness, clarity and freedom from legal jargon all too rare in many laws, the law governing sex trade in New Zealand deserves to be mentioned here (Jago 2003).

In practice, governments have developed two basic legal frameworks concerning the sex trade – criminalisation and regulation. Through criminalisation, laws forbid certain activities related to payment for commercial sexual services rather than paid sex trade itself. These activities include soliciting for clients, advertising, living off the earnings of CSWs, recruiting sex workers or helping them to circulate from one country to another. This is the most common legal framework for sex trade throughout Eastern Europe, South East Asia and the Pacific, Latin America and Africa (Launer 1993; Afrol 2003). On the other hand, regulation exempts sex industry undertakings which comply with certain conditions from criminal law. Examples of countries which have opted for this form of legal framework are Greece, the Netherlands, Germany, the state of Nevada in the US, and New Zealand (Launer 1993; Burton 1995).

These legal frameworks also target sex trade activities involving minors. From an international perspective, the ILO has put in place specific measures (Convention 182) to deal with all forms of child labour, including child sex tourism (see also Chapter 1). Due to its far-reaching effects and its enormous potential for expansion in Africa and in Kenya in particular, we end this section by examining the international reaction to child sex tourism.

International Response to Child Sex Tourism

In 1996, the World Congress on Child Sexual Exploitation was held in Stockholm, Sweden (ECPAT 2002; AVERT.ORG 2004). During this meeting, 122 governments adopted a Declaration and Agenda for Action to eliminate commercial sexual exploitation of children. Through this declaration, individual countries were required to develop a National Plan to develop local strategies aimed at dealing with all forms of child abuse (ECPAT 2002; Afrol 2003; Enberg 2008). It also required active participation of UN-affiliated bodies like the United Nations Children's Fund (UNICEF), the ILO, International Police (Interpol) and the World Tourism Organisation (WTO) (Gitonga and Anyangu 2008).

Moreover, the majority of countries are signatories of the UN's Convention on the Rights of the Child. According to Article 34 of this Convention:

> States Parties undertake to protect children from all forms of sexual exploitation. For these purposes, States Parties shall in particular take appropriate national, bilateral and multilateral measures to prevent:
>
> (a) The inducement or coercion of a child to engage in any unlawful sexual activity;
>
> (b) The exploitative use of children in prostitution or other unlawful sexual practices, and in pornographic performances. (ECPAT 2002: 3)

Consequently, many countries have passed legislation that criminalises sexual exploitation of children. Kenya, for instance, ratified the Convention on the Rights of the Child in July 1990 (Kenya 1991). In order to bring its legislation into conformity with international obligations, Kenya's parliament enacted the

Children's Act in 2002 (Kenya 2002a, 2004; Gitonga and Anyangu 2008). The Act defines a child as anyone who is under the age of 18. From a general viewpoint, this law protects children from all forms of sexual exploitation – child labour, sex trade and pornography. It also includes provisions guaranteeing free basic education and the right to health care for children. Further, the Act has established the National Council for Children's Services and the Children's Courts, all aimed at promotion, and indeed protection, of children's welfare (Gitonga and Anyangu 2008). Other legislation that protects children in Kenya includes the Criminal Law (Amendment) Act of July 2003 and the Penal Code. The Criminal Law (Amendment) Act harmonised penalties for rape and defilement to a maximum sentence of life imprisonment (Kenya 2006; Gitonga and Anyangu 2008). Attempted rape is also punishable by life imprisonment. The legal age of consent for sexual activity for girls is set at 16 years, with no minimum age for boys, while the legal age for marriage for both girls and boys is 18 years (Kenya 2006). This clause on age for marriage is intended to prohibit child marriages. On the other hand, the Penal Code covers general sexual exploitation of children (Jackson 1986). However, there are no specific anti-trafficking laws in the country (see also Chapter 1).

Internationally, a number of countries have laws that criminalise child sex tourism. Of note is the US Federal Statute which has an extra-territorial law against children's sexual exploitation (Truong 1983; Jallow 2004; Kibicho 2004a). Title 18, Section 2423 of this law states: 'it is a crime for any American citizen to travel abroad with the intent to sexually abuse children' (ECPAT 2002: 1). Under the provisions of this Act, the American Immigration and Customs Enforcement (AICE) unit is mandated to prosecute sexual offenders irrespective of where the crimes are committed. With respect to this law, an American (name withheld for anonymity reasons) aged 50 years was jailed for 25 years by an Oregon court in April 2005 for producing child pornographic films in Kenya. He had filmed himself having sex with a 10-year-old girl, the daughter of a Kenyan woman he had married in June 2004 (Kithaka 2004). Going by this conviction, the message is clear – any American who sexually exploits children should no longer expect distance or the anonymity of cyberspace to shield him or her from justice (ECPAT 2002; Sarpong 2002).

Likewise, Australia and Italy have gone a step further and designed laws which are more in tune with the ever-changing tourism industry (ECPAT 2002; Afrol 2003). In Australia, for example, it is possible to prosecute tour operators and travel agents who 'encourage' child sex tourism (Rao 2003). Such encouragement ranges from organising sex tours within and outside Australia with consumption of child sexual services as the main or part of the attractants to marketing such services (Sarpong 2002; Afrol 2003). On the other hand, the Italian laws require all travel businesses to display a warning on their brochures, ticket wallets, baggage labels, in-flight videos and education packs detailing the dangers of exploiting children sexually. They also cover legal punitive measures against offenders (ECPAT 2002; Sarpong 2002; Afrol 2003).

In addition to the establishment of laws, many governments are working together to design bilateral and multilateral agreements on exchanging intelligence and police training on issues related to child sexual exploitation. Further, these agreements are intended to make sure that international borders are no barriers to the enforcement of child exploitation laws (Afrol 2003). However, actions by some governments may directly or indirectly encourage child sex tourism. National governments in countries which are struggling economically have become increasingly tourist-oriented in their search for profitable sources of foreign exchange (see, for example, Chapter 3) (Turner and Ash 1975; Krippendorf 1977; De Kadt 1979; Hall 1994; Wilson 1997; Weaver 1998; WTO 2004). Sometimes these governments ignore the effects of the sex industry in order to encourage overall tourism development in their countries (Cabezas 1999; Ryan and Hall 2001; Sarpong 2002; Afrol 2003; Jallow 2004). This directly or indirectly allows the sex industry to perpetuate sexual exploitation of children (Ndune 1996; Afrol 2003; Jallow 2004; Kithaka 2004; Gitonga and Anyangu 2008). Moreover, in most cases child exploitation laws do not specify the prosecution procedures. For example, the laws fail to answer the following questions: Who is to file the complaint? How will the evidence be gathered? How are the children concerned going to testify in a foreign country, and at times in a foreign language? These questions need to be addressed in order to make the legislation effective.

Sex Trade and the Law in Kenya

As in many other countries, the legal situation of sex trade in Kenya is marked by ambiguities and confusion. According to Kibicho (2004a: 190), Chapter 63, Sections 153 and 154 of Kenya's Constitution render sex trade illegal with the following words:

Section 153: (1) Every male person who –

(a) knowingly lives wholly or in part on the earnings of prostitution; or

(b) in any public place persistently solicits or importunes for immoral purposes, is guilty of misdemeanour; and in the case of a second or subsequent conviction under this section the court may, in addition to any term of imprisonment awarded, sentence the offender to corporal punishment.

(2) Where a male person is proved to live with or to be habitually in the company of a prostitute or is proved to have exercised control, direction or influence over the movements of a prostitute in such a manner as to show that he is aiding, abetting or compelling her prostitution with any other person, or generally, he shall unless he satisfies the court to the contrary be deemed to be knowingly living on the earnings of prostitution.

Section 154: Every woman who knowingly lives wholly or in part on the earnings of prostitution, or who is proved to have, for the purpose of gain, exercised control, direction or influence over the movements of a prostitute in such a manner as to show that she is aiding, abetting or compelling her prostitution with any person, or generally, is guilty of a misdemeanor.[2]

A strict interpretation of the above law is that there is no illegality in sex trade *per se*, with criminality only lying in living off earnings from it. For instance, one is tempted to ask, is a woman who offers sexual services, and in return, instead of asking for payment, uses synonyms recognisable by both parties in the name of 'appreciating for services offered', contravening this law (see Chapters 6, 8 and 9)? The law fails to define clearly who a CSW is. Also, it does not show what the phenomenon of sex trade involves. It is probably due to this ambiguity on the part of the law that in many instances when law enforcers arrest young women in the urban centres in the country, they are never accused of practising prostitution. Instead, they are charged with either loitering or vagrancy, or both (Backwesegha 1982; Jackson 1986; Sindiga 1995; Ndune 1996; Kithaka 2004). The results of this ambiguity and clear double moral standard is that, customers of CSWs – for example, pimps and the proprietors of entertainment establishments (from where CSWs operate) – are rarely ever stigmatised and seldom punished for their usage of sex workers' services. It is only the CSWs who have to bear the brunt of society's moral attitudes and enduring social and legal pressures (see also Chapter 7).

To ease this ambiguity, this study adopted Gaghon's (1968) definition of the sex trade as 'any business transaction involving granting of sexual access on a relatively indiscriminate basis for payment either in money or goods or both' (cited in Cohen 1982: 413) (see also Chapter 1). The payment should, however, be distinguishable as being for a specific sexual act. Such a transfer of money or goods is hard to classify. To validate the foregoing, consider the following argument by a tourist:

It was not sex trade in that sense, because there has been in no way and at no moment any real financial claim or expectation on part of the girls towards us. It wasn't that way, that the girls would have been financed by us, no, and it has definitely not been in that way, that they were honored by us or that they were paid in one way or another. There were really friendly relations between us ....
(Günther 1998: 73)

According to this tourist, the absence of payment shows that they had no commercial relationships with professional CSWs, but rather personal, 'friendly reactions' with two young girls. These are in fact strong arguments against classification of the two visitors as sex tourists and the two girls as CSWs. In most cases, CSWs use this 'soft-payment-technique' as a way to protect themselves from moral and legal

---

2   See also Jackson (1986: 67).

constraints (see, for example, Chapters 6–9) (Günther 1998; Ryan and Hall 2001; Kibicho 2004b). In other cases, clients state that their involvement with CSWs is sex without commitment, thrill, compensation for a sterile marriage or sexual relief. Sex workers, on the other hand, refer to their interactions with clients as business relationships (see Chapters 8 and 9). For those who consider sex trade as a business talk about 'doing body work', in which emotions are completely separated from this physical embodied experience (see also Emma's comments cited in Chapter 9). However, the bottom line is that few tourists would define themselves as sex tourists even though they fit the defined criteria (see also Chapter 5).

Sex Trade Laws and the Reality

Is sex trade a crime? Basically, a crime must have a victim (Jackson 1986; Cabezas 1999). Although moralists may argue that society as a whole is the victim, an escort agency caters to willing clients by connecting them with willing (sexual) service providers (see Chapter 4). As long as there is a demand, there will always be a supply – a basic theory of economics (Boissevain 1974; Dahles 1999; Enberg 2008). So far, we have seen a considerable debate 'for' and 'against' decriminalisation of the sex trade. It is important, however, to take an objective and indeed a realistic approach when examining laws governing the sex industry in Kenya, in Africa, and in fact in the world.

To deal with the problems associated with sex trade such as human trafficking, violence against CSWs, drug trafficking and general crime, two approaches are generally promoted. While advocates for sex trade as a form of employment suggest that legalisation will address these problems (Naibavu and Schutz 1974; McLeod 1982; Ndune 1996), conservative policymakers tend to recommend penalties as a means to deter the trade (Finnegan 1979; Robinson 1989; Findley and Williams 1991; Pickering and Wilkins 1993). These views form the heart of the discussion in the remainder of this section.

Criminalising sex trade leads to part of the industry operating underground, thus making it difficult to educate CSWs about safe sex (see Leslie's comments quoted earlier in this chapter). Middleton (1993: 61) adds: 'if the HIV-AIDS pandemic is to be reduced in any given society, sex trade must be brought out of hiding'. From an academic standpoint, therefore, if CSWs' work is considered to be contributing to societal wellbeing, rather than as a crime, it can be reasoned that more of these sex workers will become conscious of health and safety issues in the sex industry at large (see, for example, Naibavu and Schutz 1974; McLeod 1982; McLintock 1992; Middleton 1993; Hall 1996; Chapman 2005). Further, decriminalising the trade gives CSWs more control of their working conditions, and eventually empowers them to break the cycle of financial dependency on third parties – pimps and brothels (Cohen 1993; Cabezas 1999; AVERT.ORG 2004).

However, like the sex business itself, decriminalisation of sex trade is not free of contradictions. For instance, some commentators argue that the cost of legalised sex trade is passed on to CSWs (see, for example, Robinson 1989; Findley and Williams 1991; Pickering and Wilkins 1993). Such costs include rental charges to the brothel owner, medical examinations and registration fees. This has a cumulative effect on the financial needs of CSWs, consequently increasing the number of sexual encounters they must have in order to make a profit. This exposes the CSW to risks like vaginal and/or anal tearing and increased drug usage in an attempt to deal with exhaustion (Truong 1983; Jago 2003; Rao 2003; Jallow 2004). In part due to these increased costs, illegal sex trade flourishes even in destinations where prostitution is legal (Ackermann and Filter 1994; Chapman 2005). This happens because sex clients seek cheaper commercial sexual services, while CSWs circumvent the legal system in an attempt to enhance their revenue-retention capacity in order to increase their profits (see Chapter 1). Nevertheless, some STCs in Kenya are advocating decriminalisation of sex trade (see Chapters 6–9) (Kibicho 2005b). If this advocacy wins the hearts of the policymakers, however, the decriminalisation process should ensure a total compliance with prescribed standards. Adherence to the set standards would enable the legal system to eliminate the 'middlemen' and the hide-and-seek ploys that make a mockery of law enforcement in the country (Kibicho 2004a, 2004b, 2007). Such a far-reaching and probably politically unpopular decision requires policymakers and law enforcers to turn from morality and puritanical perspectives when dealing with the sex trade phenomenon. In this way, Kenya's government will spare some of its resources currently being devoted to organising ineffective crackdowns on CSWs (see Chapter 7) (Kibicho 2004a, 2004b, 2005b, 2007). As a result, the government will be able to channel its energies towards the priority societal issues such as inadequate food supply, lack of medicines in health centres, a mediocre public education system, general insecurity and chronic poverty which touch all spheres of Kenyan society.

To take care of those with moralistic tendencies, and they are many, the Kenyan government and probably other African states need to focus on a two-step strategy: (1) increased penalties for CSWs' clients, and (2) increased economic empowerment of sex workers (see also Chapter 6). To date, through enforcing the existing laws, which only target the supply side of the sex industry, Kenya's government has allowed this demand-driven business to flourish (see Leslie's comments noted earlier in this chapter) (Kibicho 2004b). Thus, there should be legal mechanisms to deal with those who sexually exploit others – boys and girls, and men and women.

Most importantly, and probably most realistically, the Kenyan government should design effective policies to address the root causes of sex trade. Such policies should help in the provision of alternative economic undertakings for the CSWs driven into the sex business by lack of employment (see Chapters 6, 8 and 9). This can be achieved through a general reduction in poverty, creation of employment opportunities, provision of micro-credit facilities, enhancing skills through training, and formulation of effective laws on child sex tourism.

Conclusion

Laws governing sex trade the world over are full of contradictions. In most countries, for example, the basic act of exchanging money for sex among adults is legal. However, most of the related activities are illegal. This makes it difficult to engage in most forms of sex trade legally. Nevertheless, Kenya's CSWs in the major urban centres and tourism destination areas have learnt how to conduct their sex business with minimal confrontations with law enforcers. In consequence, Kenya's sex industry has expanded in tandem with tourism development in the leading destination areas – the coastal region and around the Protected Areas. As in many countries where tourism is a key economic activity, however, Kenya's tourism industry was not developed with sexual services as integral part of the tourist product. In order to understand the origin of this booming industry and how it became inextricably intertwined with the sex business, Chapter 3 is dedicated to tourism development in Kenya.

Chapter 3
Tourism Development in Kenya

Introduction

Kenya lies along the east coast of Africa, covering an area of 582,646 square kilometres. The country, straddling the equator, lies between parallels 4 degrees south and north. The south-easterly shores are washed by the Indian Ocean, while to the east Kenya shares the waters of Lake Victoria, the second largest freshwater lake in the world, with Uganda and Tanzania (see Figure 3.1). Kenya's landscape varies from low-lying coastlands to snow-capped Mount Kenya – 5,199 metres above sea level. With such a contrasting topography, the country enjoys a 'varied' climate. The humid and hot coastal belt greatly contrasts the highlands. The Lake Victoria region has tropical storms, while the north has a typical desert climate. Consequently, the country's vegetation varies from dense tropical forests along the coast, to shrubbery of the arid desert lands of the north to thick mountainous forests and alpine vegetation along the slopes of Mount Kenya. This great contrast is complemented by an equally diverse wildlife – the cornerstone of wildlife-based tourism, otherwise known as safari tourism.

In 2006, the country had an estimated population of 42.7 million (Kenya 2006). Agriculture is the mainstay of the economy (Kenya 2003, 2004, 2006). Currently, tourism is the second largest contributor to the national revenues after the combined foreign exchange earnings from tea and coffee (Kenya 2006; Kibicho 2007). Kenya's tourism industry can be divided into four major forms: beach, safari, business/conference and cultural. From a general point of view, the national tourism industry has experienced a spectacular growth, especially during the post-independence era. Its potentials were recognised in the first independent Kenyan National Development Plan (1966/67–1972/73). The setting up of the Kenya Tourist Development Corporation (KTDC) in 1965 marked this recognition, not to mention the establishment of the Ministry of Tourism and Wildlife a year later. Successive economic development plans since 1972/73 laid emphasis on the industry as an engine for socio-economic development of the country.

The objective of this chapter is to analyse Kenya's tourism institutional structure and to point out the main internal and external factors that have influenced the national tourism development process from small-scale locally controlled to large-scale foreign-oriented mass tourism. It applies a longitudinal perspective of the influence of tourism on the growth of sex trade. Different stages of continuous tourism change can be captured comprehensively by the tourism area life cycle model (Butler 1980). Consequently, the model is employed

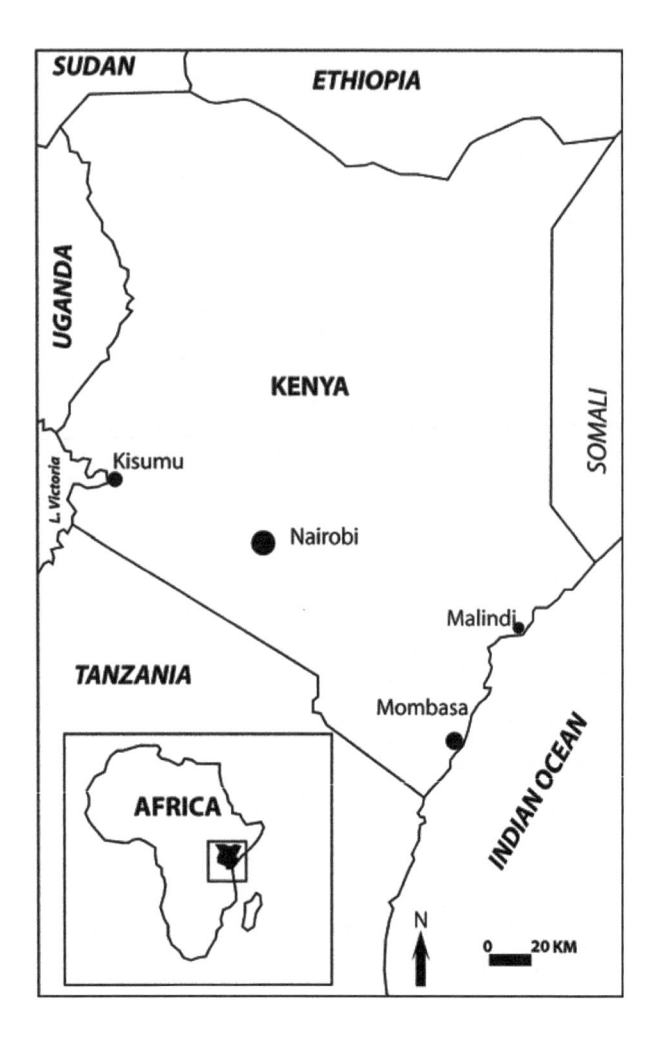

Figure 3.1   Location map of Kenya

in this chapter to explain the movement towards unsustainability of a tourism destination in its maturity stage. This approach enables us to demonstrate how unplanned and poorly managed tourism development could lead to unabated growth of tourism-related sex trade. Thus, the chapter evaluates the linkage between Kenya's tourism industry and sex trade from exploration through consolidation phases. Therefore, it plays a critical role in laying a concrete base for the discussion in the remainder of the text. Figure 3.1 shows the locations discussed.

Kenya's Tourism: Background and Economic Issues

Why Tourism Development?

At independence in 1963, the Kenyan government inherited a colonial economy that was characterised by inequitable distribution of resources, high rates of unemployment and poor living standards of the indigenous people (Kenya 1967; Bachmann 1988; Akama 1999). Furthermore, most of the country's economic activities were controlled mainly by expatriates, who had relatively high standards of living compared with those of locals. Although the initial development of the tourism industry had started in the late 1910s, when Kenya (British East Africa), became a popular destination for hunting safaris by Europeans and North Americans, there was minimal local people involvement in its development. The initial tourism facilities, for example, were developed by the resident Europeans and the colonial government. Kenyans of African origin were only hired to work in servile positions as security guards, gardeners, porters, cleaners, waiters, entertainers and so on. It was a textbook case of a master–servant relationship (see also Chapters 6 and 9, and Figure 9.2). From a general viewpoint, the initial development of tourism in Kenya was colonial in orientation, serving the expatriates' socio-economic interests (Sinclair 1990; Dieke 1991; Akama 1999; Sindiga 1999).

Consequently, the post-independence government set 'decolonisation' of the national economy as one of its principal goals (Kenya 1967). This was aimed at promoting accelerated socio-economic development through reduction of unemployment levels, alleviation of poverty and minimisation of capital deficiency. However, the government was faced with a challenge, as the national economy was heavily dependent on only one source of foreign exchange – agriculture. This economic situation further worsened in the mid-1960s, when prices of agricultural products in the world market fell drastically (Kenya 1989; Sindiga 1995). It was within this economic context that Kenya's government realised that the country had an existing 'commodity' that could create employment opportunities and generate the much-needed foreign exchange. This alternative commodity was the country's unique tourism resources – wildlife heritage (within the PAs) and pristine beaches (along the coast). Kenya's government thus embraced tourism as a tool for socio-economic development (Kenya 1967). In the longer term, it was hoped that tourism would reduce overdependency on the export of raw materials such as tea and coffee. In addition, tourism was favoured due to the multiplier effects it creates at the place of production – the tourism destination. As in other economic sectors, the government undertook various initiatives that were aimed at encouraging national tourism development.

These measures targeted both the supply and demand sides of the industry (Bachmann 1988; Sinclair 1990; Sindiga 1995). It encouraged local entrepreneurs as well as foreign investors by means of a policy that attracted foreign capital through various fiscal incentives to encourage the development of tourist

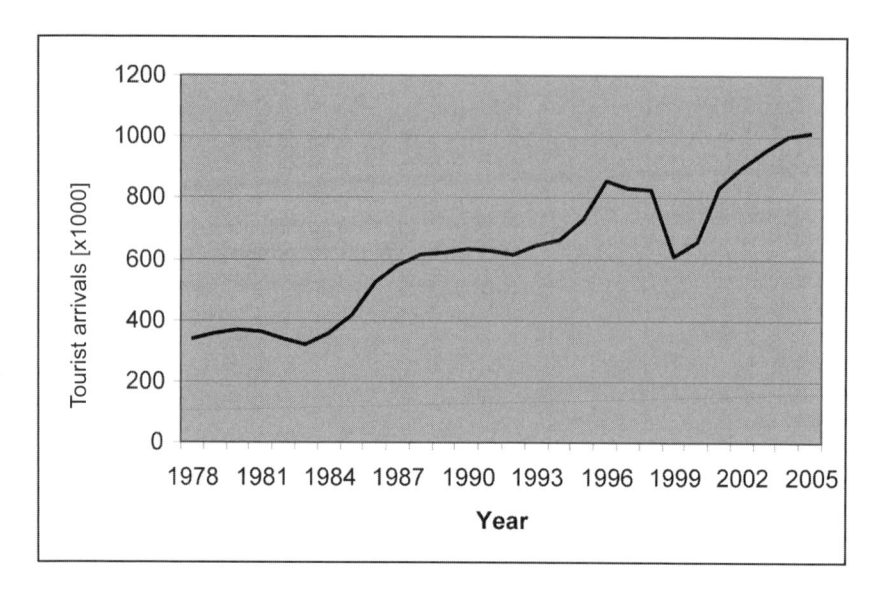

Figure 3.2 Evolution of tourist arrivals in Kenya (1978–2005)
Source: Kenya (1989; 1991; 1998; 2000; 2002; 2004; 2006)

infrastructures and superstructures in the country (Kenya 1967; KWS 1995). Foreign tourism investors were able to import equipment and personnel tax-free, and to repatriate almost 100 per cent of profits (Bachmann 1988; PTLC 1998; Sindiga 1999; Kibicho 2003). These encouraging measures favoured the national growth of tourism. International tourist arrivals, for instance, increased from a mere 35,000 in 1960 to 1,016,000 in 2005 (Figure 3.2) (Kenya 1967, 2006; Bachmann 1988; Dieke 1991; Kibicho 2007). Accordingly, the sector generated continued growth of national revenues, thereby playing an important role in the national economy in recent decades (Kibicho 2007). Thus, by the mid-1980s and early 1990s, Kenya had emerged as a leading long-haul destination in eastern Africa, receiving over 7 per cent of the total arrivals in Africa (Bachmann 1988; Sinclair 1990; Dieke 1991; KWS 1995; Weaver 1998; Sindiga 1999; Akama 2002; Kibicho 2003). In addition, the country's total tourism revenues increased from K£7.2 million in 1963 to K£28.3 billion in 2005 (Kenya 1967, 2006). In fact, in the late 1980s the country's total tourism revenues surpassed, for the first time, combined earnings from Kenya's leading export crops, tea and coffee (see Table 3.1) (Kibicho 2006b, 2006d). Accordingly, in 2005 the industry accounted for over 12 per cent of the country's GDP. It also provided 400,000 direct jobs and an estimated 550,000 indirect jobs nationwide (Kenya 2006; Kibicho 2006b).

As in other developing countries, international tourists dominate Kenya's tourism industry. The majority of the tourists to Kenya come from the UK (33

Table 3.1 National earnings from leading export crops and tourism (in K£ billions[a])

	1987	1988	1989	1990	1991	2001	2002	2003	2004	2005
Leading export crops	0.42	0.52	0.53	0.57	0.61	1.27	1.34	1.44	1.80	1.94
Tourism	0.32	0.38	0.47	0.58	0.64	1.22	1.18	1.29	1.91	2.45

Note: [a] US$1 = K£3.8.
Source: Kenya (1993, 2002a, 2004, 2006).

per cent), followed by Germany (28 per cent), the US (26 per cent) and Israel (7 per cent) (Kenya 2004). In 2006, Europe generated 68 per cent of the total tourist arrivals in the country (Kenya 2006). Specifically, Germans are the majority in beach tourism (49 per cent), Americans dominate safari tourism (42 per cent), while the British lead in business tourism (45 per cent) (Kenya 2004, 2006). Kenya's tourism industry is highly seasonal, with peak and off-peak seasons. The peak season is November–March, while the off-peak season is April–July; the August–October period could be described as 'mid-season' (see also Chapter 6). Tropical beaches on the coast support beach tourism, while inland PAs serve as the backbone of safari tourism.

Framework of Analysis

Kenya's tourism development can be divided into four major stages: (1) exploration (pre-independence period), (2) involvement (1963–78), (3) development (1979–96), and (4) consolidation (1997–2005) (see Figures 3.3 and 3.4) (see also Butler 1980). The exploration stage is signified by a small number of tourists, lack of specific facilities for tourists, and visitors being attracted by the destination's natural uniqueness. Tourist arrivals increase during the involvement stage. In addition, primary tourist facilities are put in place, and destination marketing strategies are being developed. The development stage is marked by intensive marketing of tourism. Local control and involvement in the industry decline as foreign-owned organisations get involved in tourist product/services provision. The growth rate in terms of tourist arrivals declines as the destination moved into the consolidation stage. Although the absolute number of tourists increases, the rate of growth number is declining. At this stage there is a reduction in quality of the local environment, both socially and physically, and discontent among the host communities towards tourists and the tourism industry in general (see Butler 1980 for a detailed description of these stages).

Number of tourists

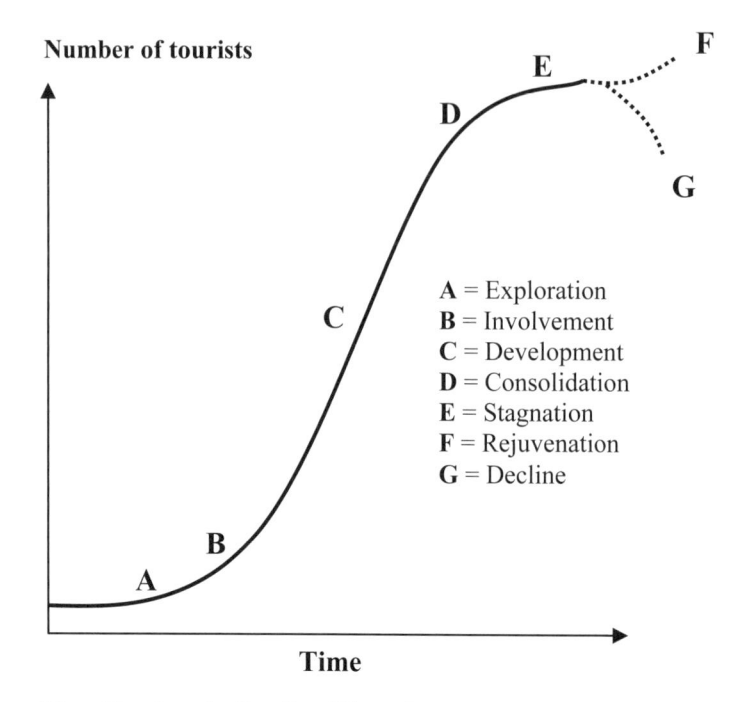

A = Exploration
B = Involvement
C = Development
D = Consolidation
E = Stagnation
F = Rejuvenation
G = Decline

Time

Figure 3.3 Tourism destination life cycle
Source: Butler (1980: 8)

Exploration Stage: Pre-independence Period

During the colonial period, Kenya received several thousands of tourists each year (Kenya 1967). They were mainly coming for game shooting safaris. The construction of the Kenya–Uganda Railway in the late 1890s popularised Nairobi as the capital for safari tourism (Wels 2000). Undoubtedly, the most elaborate hunting safari was the one undertaken by Theodore Roosevelt, the 26th president of the USA, in 1909. When the Roosevelt's train arrived at Kapiti Railway Station, there was no possible doubt about the importance of this safari. Many porters were mobilised to transport the hunting gear through the savannah: saises,[1] gun bearers, tent men, 46 Americans in charge of the weapon cases and tents, 15 police officers in uniform and 265 Kenyan carriers, enthusiastic and facetious in khaki shorts, blue puttees and blue jerseys. There were eight large tents for the hunting party, a mess tent, a skinning tent and 15 small tents for the Kenyans, all ranked behind a large US flag (Kenya 1967; KWS 1990; Wels 2000). Many animals were killed and a lot of game trophies were collected during the trip. Consequently, Kenya

---

1 Those who skin and then stuff the wild animals killed during hunting safaris in Africa.

was placed in a good position on the international map as a destination for game shooting safaris (Sinclair 1990; Wels 2000). In 1910 and 1911, for example, 715 safaris were organised. During these trips, safari hunters killed 851 lions, 995 rhinoceros and several thousands of other animal species (KWS 1990).

Hunting safaris were highly romanticised, and played a crucial role in the construction of a masculine and 'being-in-power' identity of white people in Africa (Wels 2000). The hunting experience was imbued with romance and identity construction in terms of hardship, being-in-charge and authenticity. This inspired in a whole range of thrilling stories of men as they should be – the hunter as role model, with a self-image of physical and mental superiority (Wels 2000). The colonial hunter was one of the most striking figures of Victorian and Edwardian times (Rutherford 1997; Wels 2000). The physical aspect of the identity was greatly 'propagated at British public schools through games like rugby where this image of masculinity was inculcated in young men' (Ryan 1997: 99). These schools came to be known as the 'nurseries of the empire ... where a collective practice of asceticism intent on transforming the feminine bodies of young boys into the hardened musculature of imperial warriors was implemented' (Rutherford 1997: 16).

Hunting seemed to be the ideal expression of this masculinity, and thus played a highly significant role within colonial expatriate culture. Wels (2000: 106) sums it up by saying that 'a safari hunter had to match his prowess as a sportsman with the skill of a diplomat. His shower water while in the safari camp had to be piping hot and his cocktails ice-cold'. The clothing the hunters wore was another expression of this masculine identity. It was not only selected for its convenience in the bush, but was also a way of expressing an image of manliness in the broader context of imperial power relations (KWS 1995). It is remarkable to note that this aspect of the hunting outfit has not changed over the years. Pictures show almost the same outfits over the decades: shorts, often with a belt to which a knife is attached, high leather hunting boots which protect the ankles, high socks which hardly ever rise up to their knee-high potential, a shirt with short or rolled-up sleeves, epaulettes and two breast pockets with an overhanging flap which can close the pocket with a small button, and a hat to complete the outfit, all in shades of khaki. Today, however, hunting is illegal in Kenya – hunting safaris have been replaced by photo-shooting safaris (safari tourism).

The relationship between hunting safaris and sex trade in Kenya, however, is not well documented. Existent literature only mentions the exotic and erotic African female companions who accompanied the hunters throughout their expeditions (see, for example, Backwesegha 1982; Sindiga 1995; Ndune 1996). Safari hunting companies in Nairobi hired these companions as 'housemaids' on behalf of the safari hunters. However, none of them was involved in 'housekeeping' chores (Ndune 1996). In fact, their roles were never specified. They served as hunters' 'concubines', and thus offered free sex access to them during the hunting safaris. The term 'concubine' is more accurately applied to the companions chosen by male colonial personnel – military men and colonial officers who spent several years in one place in an era when European wives seldom went to Africa (Sindiga 1995; 1999; Ndune 1996; Rutherford 1997; Wels 2000). Thus, to keep an African woman was 'a practice accepted and even recommended by military doctors, who saw it as a kind of health insurance' (Sindiga 1995: 3). These women were to 'amuse', care for, dispel boredom and keep the European man from turning to alcoholism and sexual depravities (see also Mensah 2005). They were above all to be beautiful and free-spirited. Young female Kambas, a Kenyan ethnic community, were particularly sought after (see, for example, Backwesegha 1982).

During the hunting safari, the hunters' concubines were also required to dance around the evening campfire naked to entertain the (whole) hunting troupe (Backwesegha 1982; Sindiga 1999). In almost all cases, female companions were introduced to the safari hunting companies by their acquaintances in the 'hunting fraternity'. Ndune (1996: 9) notes: 'these women knew that they could improve themselves, and enhance the prospects of their families, by relations with white safari hunters'.[2] In fact, a number of them became modestly prosperous from their relationship with the hunters (Backwesegha 1982; Sindiga 1995; Ndune 1996). It enabled the more nimble and creative African women to acquire a host of luxury goods and hard cash to start small-scale enterprises. Other things being equal, it can be argued that at this stage of Kenya's tourism evolution, the linkage between tourism and sex trade was slight, if not insignificant (see Figure 3.4).

Involvement Stage: 1963–78

As noted above, immediately after independence, Kenya's government focused on national socio-economic growth. It thus assumed that:

1. direct foreign investments would solve the bottleneck problem of scarce domestic capital;
2. foreign exchange earnings would help to reduce notorious balance-of-payment deficits;
3. employment generation would mitigate social inequalities;
4. multiplier effects in peripheral areas would lessen regional disparities (see Kenya 1967).

Towards this end, the national tourism strategy was to increase foreign exchange earnings and to generate employment opportunities through increased tourist numbers (Kenya 1991). Tourism development was therefore institutionalised.

---

2   This is a typical African colonial mentality – see also Chapter 6.

Establishment of the KTDC and the First Tourism Ministry

To promote rapid expansion of national tourism, the Kenyan government established the KTDC (a quasi-governmental organisation) in 1965 to oversee tourism development and operation of tourism facilities. The corporation is also involved in granting loans, and attracting local and international investors, not to mention other administrative duties. In 1966, for instance, the KTDC acquired tourism investment funds from bilateral and multilateral financial institutions on behalf of the government of Kenya (Kenya 1967; 1989; Bachmann 1988; Dieke 1991; Weaver 1998; Akama 1999, 2002). These funds enabled the corporation to undertake intensive capital investment in developing tourism facilities during the first phase of this incipient stage of the Kenya's tourism development (mid-1960s to mid-1970s).

Some of tourist facilities that were established with these funds are: hotels, including Buffalo Springs, Golden Beach Resort, Kabarnet, Kilanguni, Milimani, Mombasa Beach Resort and Sunset, and lodges, including Lake Bogoria, Mountain Lodge, Ngulia, Olkuruk Mara and Voi Safari. These hotels and lodges were managed by the now defunct African Tours and Hotels (AT&H), a quasi-autonomous government organisation established in 1974 to work in collaboration with the KTDC. This organisation was abolished in 1998 due to financial losses resulting from large-scale mismanagement and gross inefficiency. Towards the end of the same year, all tourist facilities hitherto managed by AT&H were under receivership owing to increased loss-making. This poor performance by AT&H, and its affiliated hotels and lodges, was largely due to political interference in their daily operations. Hiring (and firing) of personnel, for instance, was strictly based on kinship ties and political affiliations. This may have not been the only reason for AT&H's underperformance, but it was certainly a principal contributory factor. This confirms the argument by Jenkins and Henry (1982) that most government entrepreneurial initiatives are usually driven by political considerations rather than economic relevancy and financial expediency.

In 1967, the International Finance Corporation (IFC), the private-lending arm of the World Bank, in collaboration with the Inter-Continental Hotel Corporation, gave the Kenyan government a concessionary loan of US$3 million to be used for tourism facilities development (Bachmann 1988; Sinclair 1990; PTLC 1998). Consequently, between 1968 and 1975, KTDC subsidised the construction and renovation of more than 50,000 hotel rooms (Sinclair 1990). This marked the beginning of continuous rapid growth in national tourism and hospitality facilities. Accordingly, by the end of this second stage (1963–1978) of Kenya's tourism development, the number of international tourists had increased more than tenfold (Kenya 1989; Sinclair 1990).

The Kenyan government's recognition of the economic importance of the tourism industry was further confirmed by the creation of a fully fledged Ministry of Tourism and Wildlife in 1966. This ministry was in charge of all matters pertaining to the development and management of the national tourism industry. More specifically, the functions of the ministry were to:

1. increase the overall contribution of tourism to the GDP through increased foreign exchange;
2. train and develop professional manpower for the tourism industry;
3. establish and manage hospitality and other tourist facilities;
4. initiate and promote private investments in the tourism industry;
5. regulate the establishment of hospitality and other tourist related facilities;
6. promote and market Kenya as a tourism destination, both locally and internationally. (Kenya 1967: 33)

Its first task was the formulation of a national tourism policy; thus the ministry formulated Sessional Paper No. 8 of 1969 on tourism development (Bachmann 1988; Sinclair 1990; KWS 1995; Kibicho 2003). The document details tourism policies for the future, directed towards the achievement of three major goals:

1. **Economic** – to minimise constraints on tourism growth, to contribute to the balance of payments and to stimulate employment growth;
2. **Environmental** – to contribute to conservation of the natural and cultural resources, and
3. **Social** – to encourage appropriate operation of tourism activity in the public interest and to support domestic tourism. (Kenya 1967, 1987; Kibicho 2003)

However, as Hall (1994: 114) observes: 'it is not just the range of objectives that needs to be considered but the relative priority attached to objectives as they are implemented'. The government of Kenya, for instance, gave economic goals far higher priority than the other two. Thus, policy implementation focused primarily on providing infrastructural support for large-scale, enclave-like projects in order to meet the demands of the ever-wealthier international clienteles. Lack of domestic capital and supportive infrastructure were considered obstacles to tourism-based economic growth (Dieke 1991). In order to overcome these limitations, the ministry initiated a tourism master plan to increase foreign capital investment through tax concessions, favourable fiscal incentives for capital investment, and profit repatriation. These measures were aimed at prompting international investors to invest in tourism projects that were otherwise too expensive due to the inadequate nature of their infrastructure, and too risky as the country lacked an image as a tourism destination (Kibicho 2003). As a product of these national tourism policy transformations, Kenya's tourism became a major socio-economic phenomenon.

However, the Ministry of Tourism and Wildlife, has never been an institution in its own right. The ambiguity and confusion over this ministry and its respective organisation was perpetuated as it was placed under the auspices of successive ministries over the last four decades: the Ministry of Tourism and Wildlife

Management, the Ministry of Tourism and Culture, the Ministry of Tourism and Natural Resources, the Ministry of Home Affairs and Tourism, the Ministry of Tourism and Environment, the Ministry of Tourism, Trade and Industries, the Ministry of Tourism and Information, and the Ministry of Tourism and Wildlife (the most recent change). This new title is evidence of the strong link between Kenya's tourism development and wildlife/PAs. From the foregoing, it is clear that as a new government came to power, tourism and its fate remained uncertain.

From Small-scale Businesses to Mega-scale Investments

During the last phase of this stage, from the mid- to late 1970s, the dependency paradigm gained ground. Increased investments by transnational companies in developing countries were seen as a strategy by the developed countries to take advantage of the resources of the global South (Britton 1982; Burns and Holden 1995; Brenner 2005). Consequently, Kenya's tourism industry changed from small-scale public and private businesses to mega-scale investments funded by overseas multinational companies and investors. Tourism policy was thus directed towards large-scale investments in co-operation with transnational enterprises.

Subsequently, tourist products and services were defined, owned, and indeed provided by these firms. As the government has become dependent on these firms for the economic development of the state, their operations are given a free reign. Orientation of administrative services, labour regulations and marketing strategies all proceed in accordance with the requirements of these dominating (in most cases) foreign-owned enterprises, at the expense of informal, small-scale, locally owned business organisations. This kind of dependency on external capital investment means that the country's tourism industry could be greatly influenced and dictated by unpredictable exogenous socio-economic and political factors. From a general point of view, the challenge here can be said to be the welter of dependencies that have crystallised: on imported skilled labour, foreign capital, imported goods, international development and planning (by transnational hotel chains), and of course, tourists themselves, largely from abroad. The concern is that the locus of control over the tourism development process shifts from the tourist host community, the people who are most affected by the development, to the tourist-generating areas (see Chapter 9). Given such external control over the fortunes of the tourism industry in Kenya, and indeed in other developing nations, it can be argued that tourism is too fragile and unpredictable an industry on which to base total economic development of a destination (Britton 1982; Cazes 1992; Dahles 1999; Hall and Tucker 2004).

During this stage, Kenya's economy was robust (Kenya 1967, 1987). Unemployment levels, for instance, stood at 6–8 per cent. Accordingly, poverty levels were relatively low (Kenya 1967). Consequently, poverty sex trade was marginal (Kibicho 2004b). However, as tourism destination areas at the coastal region and around PAs evolved, so did the number of CSWs targeting international tourists as their clients (see Figure 3.4) (Kibicho 2004b, 2006a). This trend was specifically visible in Malindi Area in the early 1970s: 'As the number of tourist arrivals increased, the number of CSWs increased proportionally,' said the MWA's secretary. She further notes:

> any extra hotel opened in this period brought in about 14 new CSWs. The strong tourism wave of the 1970s attracted many Kenyans to Malindi. Those who did not have the required skills to work in the industry, like my late mother, ended up in the sex business. As a result, Malindi's sex industry grew in tandem with tourism expansion. For sure, tourism supported sex business, but I am not certain to what extent sex trade influenced tourism activities in that period.

Development Stage: 1979–96

The first phase of the development stage of Kenya's tourism development was marked by the imposition of the Structural Adjustment Programme (SAP)[3] on Kenya and other developing nations by the World Bank and the International Monetary Fund (IMF). Consequently, many bilateral and multilateral institutions gradually retreated from offering direct support to Kenya's government (Kenya 1989, 1990; Sinclair 1990; Weaver 1998). None the less, the government continued to support the KTDC activities mentioned earlier, albeit with reduced financial means. Subsequently, most state-owned tourism enterprises (until then performing poorly) were privatised. From a macro-perspective, this privatisation process was (and still is) a concrete example of 'institutionalised corruption', as bidders were only considered on the basis of their relationships with the ruling political class (the 'big men') – political cronyism (Kibicho 2007). This was mainly due to an absence of institutional and legal mechanisms to guide the privatisation process. The notion of dependency discussed in the previous section can thus be linked to the role of these local elites' vested interests as they control the privatisation process, sometimes in conjunction with foreign investors' interests (see, for example, Hall and Tucker 1994; Kibicho 2003; Hall 2004).

This new form of 'dependency' involved subordination of national tourism autonomy to meet the interests of privileged local classes and foreign investors, rather than those development priorities arising from a broader consensus. The local elites–foreign investors relationship and the pursuit of tourism development objectives by local elites which are exogenously derived is a clear manifestation of the 'colonial plantation economy' in the national tourism industry (Kibicho 2003; Hall 2004). Thus, the overall development is similar to colonial commodity fiat, wherein a colonising European power would define an arbitrary price for natural resources, extracting which diminished national natural capital. Hall (1994: 127)

---

3   The structural adjustment policies (a) let market forces set relative prices, (b) cut back state expenditure and intervention, and (c) liberalised the economies and opened them up to international trade and foreign investment.

elucidates the foregoing by saying that: 'tourism in many developing countries ... exhibits the characteristics of a colonial plantation economy in which metropolitan capitalistic countries try to dominate foreign tourism markets'. In this kind of tourism development, overseas interests are the determining factor in the creation of both the demand and supply of tourist products and services. It can thus be said that the processes occurring at the micro-level (through examination of the interests, values and power of significant local individuals and groups) and macro-level (domination by foreign enterprises) should not be regarded in isolation; but instead should be seen as being entwined with broader global capitalistic patterns of tourism development.

By the early 1980s, Kenya's tourism development was characterised by unbroken continuous growth, mainly through the initiatives of foreign investors (Kenya 1987; Bachmann 1988; Sinclair 1990; Dieke 1991). It was a phase of 'tourism industrialisation', where the tour operator controlled the whole of the tourist product production process, including chartered flights, accommodation and transport while in the visited destination – all-inclusive tourism packages (we will return to examine the effects of this form of tourism on sex trade growth and destinations' sustainability in Chapter 9). During this period, the demand side of the tourism industry largely exceeded that of the supply. Consequently, the tourism resource base (for both safari and beach tourism) was overwhelmed by haphazard and uncoordinated tourism development (Weaver 1998; Akama 1999, 2002). The government's aim was to meet the demands of the fast-growing mass tourism sector. Up to the present, the government's policy is to enhance tourism revenues and to encourage higher tourist arrivals, mainly by providing the necessary facilities, on the assumption that visitor arrivals will continue to increase every year, the essence being to 'break records' at any price. The industry is thus guided by a *laissez-faire* policy. As a result, to a greater extent, Kenya's tourism corresponds to a model which one could describe as mass tourism well marked by a wide range of effects, including expansion of tourism-related sex trade (see Chapter 1) (Kibicho 2004b, 2005a, 2005b, 2006c, 2007).

Tourism Policy Modifications

Tourism development and environment conservation policies in Kenya are guided by the government's need to generate revenue (KWS 1995; Weaver 1998). Therefore, there is a strong link between safari tourism and the existence of the PAs which cover about 8 per cent of the country's total surface area. Safari tourism depends heavily on these conservation areas (Bachmann 1988). However, there are unending conflicts arising from the existence of the PAs and other forms of land use such as mining, agriculture, pastoralism and forestry (KWS 1995). According to Runte (1987), safari tourism – and hence the existence of the PAs – reduces agricultural productivity in the pastoral areas. This is because the PAs occupy most of the areas with relatively high agricultural potential. Apart from land use conflicts, there will always be positive and negative, real and perceived, and direct and indirect effects associated with tourism development.

After 1985, noteworthy policy modifications in tourism development took place. Centralised state interventionism continued, but the concept of wildlife-based tourism gained a certain importance (KWS 1994, 1995). This was marked by the creation of the Kenya Wildlife Service (KWS), which replaced the Wildlife Conservation and Management Department (WCMD) in 1989. To understand the reasons behind this institutional change, let us look briefly at the historical background of wildlife conservation in Kenya.

The first legislation on wildlife was enacted in 1898 (KWS 1990, 1994, 1995). This legislation established game reserves and introduced controls on hunting. In 1907 a Game Department was established to manage wildlife throughout the country. Ordinance 9 of 1945 established a Board of Trustees to administer land set aside as PAs. In 1976, the functions of the Game Department and the National Parks were amalgamated under the WCMD (Kenya 1990). The record of this department's operations was disappointing, mainly because of lack of funds (Kenya 1987; KWS 1995). Reduction of public funding was mainly because in the early 1980s, the government implemented a programme of economic stabilisation measures which led, among other things, to decline in real financial allocations to the WCMD (KWS 1990, 1995, 1997). As a consequence, the recurrent budget in 1980/81 and 1981/82 averaged K£47.1 million. In 1988/89 and 1989/90 it was K£72.5 million per year, a substantial drop in real terms (KWS 1990). The shortage of operating funds led to poor management standards. Low salaries and failure to pay the allowances due caused staff demoralisation. In the late 1970s, corruption increased in the wildlife sector. Senior wildlife officers took an active part in poaching, which became practically uncontrollable by the end of 1988 (KWS 1990, 1997; Weaver 1998). The populations of elephants and rhinoceros, for instance, reduced by 85 per cent and 97 per cent respectively (see, for example, Kenya 1987, 1989; KWS 1997; Weaver 1998; Kibicho 2005c).

Meanwhile, as noted above, tourism had grown through the 1980s, and its economic importance was unparalleled. This growth partially concealed the developing crisis in wildlife management which was by then jeopardising the future of the tourism industry. This was manifested by an inability to guarantee the safety of tourists and their growing dissatisfaction with poor tourist infrastructures and superstructures within and outside the PAs. In an attempt to address this problem, the government made changes in the leadership of the conservation department. These institutional reorganisations were crowned by the establishment of the KWS in place of the WCMD. Its specific objectives are to:

1. conserve natural environments and their flora and fauna;
2. use wildlife resources sustainably for the economic development of the nation and for the benefit of people living in wildlife areas;
3. protect people and property from injury or damage caused by wildlife. (KWS 1990; Weaver 1998)

Kenya's wildlife policy is intimately linked with tourism and development of rural areas. It revolves around the idea of sustainable utilisation of natural resources to generate revenue through a series of activities, notably tourism (KWS 1994). The KWS Policy Plan makes it clear that tourism is perceived as the key to the economic survival of the PAs. It mitigates some of the very high costs of protection and maintenance of PAs in Kenya (KWS 1990). Like many other of Kenya's public institutions, the KWS suffers from inadequate governmental funding (KWS 1990; Dieke 1991; Kenya 1996, 1998; Sindiga 1999). This has reverberations throughout the PA system, as projects are reduced to cut costs. In addition, park personnel are underpaid and ill-equipped, leading to staff demotivation. The major part of this problem is that only a small portion of parks' revenues is ploughed back into the PA system (Bachmann 1988; KWS 1990).

To compensate for the government shortfall, although still not adequate to meet all the requirements of the PA system, the KWS receives financial assistance from diverse sources. These funds come from international agencies (such as the African Conservation Centre, African Wildlife Fund and World Wide Fund), multilateral agencies (for example, the Global Environmental Facility of the World Bank), international development banks and individual/private donors (KWS 1990; Kenya 1996). In general, this kind of funding leads to three principal problems:

1. lack of co-ordination among the participating organisations, resulting in redundancy;
2. loss of control and power by the KWS as donors dictate how the money should be used, which leads to a dependency relationship;
3. the availability of alternative sources of funding might have acted as a disincentive for the government to increase its own contribution to the PA system (see also Weaver 1998).

Another significant policy change was the development of the latest national tourism master plan in 1995. This ten-year tourism development plan (1995–2005) was undertaken by a consortium of Japanese consultants, and sponsored by the Japanese International Co-operation Agency (JICA).

The master plan is critical of the lack of achievement in tourist product diversification, and it states: 'regardless of efforts to diversify the national tourist product, many regions with tourism development potentials have not been exploited' (cited by Kibicho 2004a: 10). This is mainly because Kenya's tourism resorts evolved as enclaves, resulting in a heavy concentration of tourism activities in a few areas. The most significant areas in terms of tourist concentration are Nairobi (as urban and conference/business tourism), coastal region (as sand-sun-sea-sex activity, especially in the Mombasa–Malindi corridor) and PAs (as safari tourism, especially in the Maasai Mara National Reserve, Tsavo West and East, Nairobi, Amboseli, Lake Nakuru and Arberdares national parks). These areas, covering about 20 per cent of the national surface area, accounted for 84 per cent of tourist accommodation in the country in 2000 (Kenya 2001). Subsequently, this tourism development plan was aimed at identifying the tourism potential of the six tourism regions: the central, coastal, eastern, northern, southern and western tourism circuits. Thus, for the first time, the government intended to address the problem of spatial distribution of tourists in the country. However, this plan and policy document has not been implemented, owing to an absence of clearly defined implementation procedures – allocation of sufficient funds, tourism expertise, and who is to do what during the process.

From a general observation, tourism policymaking in Kenya is highly centralised, involving mainly the top government officials and the services of foreign consultants (Akama 1999; Kibicho 2007). Consequently, there is no participation from other stakeholders, especially local communities, local authorities and the private sector in such initiatives. It should be noted that 'without agreed aims and objectives, *by all the interested parties*, formal development planning is likely to be uncoordinated and unsatisfactory' (Lickorish et al. 1991: 67; my emphasis). In addition, such a process should be flexible, dynamic and evolutionary, which involves continuous evaluation of various aspects of tourism development – demand and supply, growth trends, and existing infrastructures and superstructures. When all is said and done, a specific institution should be responsible for policy and implementation of plans.

These tourism policy modifications, however, did not envisage possible linkage between tourism expansion and sex trade growth. In all cases, sex issues were marginalised in an attempt to portray the country as a family-friendly long-haul tourism destination. However, the reality is that during this stage, especially in the early 1990s, Kenya became a textbook case of a cheap (mass) tourism destination (Akama 1999; Sindiga 1999; Kibicho 2003). Tourists of all descriptions flocked to the country. Expanding poverty levels compounded by lack of employment opportunities due to a stagnating national economy led many Kenyans to migrate to key tourism destination areas in search of elusive jobs. As mentioned in Chapters 1 and 6, many of these immigrants were not absorbed into the mainstream tourism industry. Some of them opted to try their luck in the sex profession (Kibicho 2004b). Of course, not all of them targeted tourists as their source of business. However, one thing is certain, as the tourism industry performed better, which meant more money in circulation in the local economies, competition among CSWs for clients eased (Sindiga 1995; Ndune 1996). This was because of an increased demand for commercial sexual services, as more locals could afford to pay for these services. Thus, one may argue that the linkage between sex trade and tourism became relatively more explicit at this stage of Kenya's tourism evolution (Figure 3.4) (see, for example, Sindiga 1995; Ndune 1996; Kibicho 2004a, 2004b, 2005b, 2005c, 2006c).

Table 3.2   Hotel bed occupancy (%)[a]

	All Kenya	Coast	Nairobi	Lodges
1985	49.6	59.6	52.7	52.2
1986	49.4	62.2	48.5	52.0
1987	50.0	58.8	50.2	54.5
1988	43.1	58.7	55.5	58.3
1989	41.0	55.0	51.0	57.8
1990	42.1	58.0	50.5	59.8
1991	40.3	60.0	53.3	47.7
1992	39.2	56.3	45.5	55.1
1993	54.8	63.3	61.5	60.5
1994	52.1	62.4	58.3	57.5
1995	63.4	66.1	65.0	66.2
1996	64.2	68.4	64.1	65.0
1997	39.0	28.3	40.6	38.3
1998	38.4	27.3	38.0	37.2
1999	40.5	30.0	41.4	41.5
2000	41.1	30.5	40.2	42.8
2001	40.6	35.7	42.9	43.2
2002	40.0	37.8	42.0	44.5

Note: [a] Occupancy rate (%) is the number of occupied bed nights divided by the yearly available number of bed nights.
Source: Kenya (1987, 1989, 1991, 1996, 2001, 2002a).

Consolidation Stage: 1997–2005

By the mid-1990s, Kenya's tourism industry was showing remarkable signs of poor performance, well marked by stagnating hotel occupancy rates (see Table 3.2). The average length of stay of visitors increased gradually from 8 to 13 days in the first two sub-periods, but dropped to 6 days in the sub-period 1990–2000. Naturally, the lower growth rate of length of stays would be reflected in room bed occupancy levels (see Table 3.2).

Table 3.2 shows that the occupancy rates declined drastically in the period 1997–2002 as a result of expanding political violence and more general economic decline (Kenya 2003, 2004; Kibicho 2007). Widespread reporting of political violence in the run-up to the 1997 elections exacerbated this tendency as international tourists shunned the country, opting to go to alternative tropical destinations. During this period, Kenya lost its tourism market share to its key competitors, notably Botswana, South Africa, Tanzania and Zimbabwe (see, for example, PTLC 1998).

Establishment of the Kenya Tourist Board

This unforeseeable and abrupt decline in the industry led to further policy modifications. The most significant one was the establishment of the Kenya Tourist Board (KTB) in 1997. The KTB was expected to promote Kenya as a tourism destination in a more co-ordinated and systematic fashion. More specifically, the KTB was supposed to:

1. produce base-line data describing key features of tourism as a benchmark for future marketing;
2. identify specific tourism market segments;
3. stimulate consumer demand for travel to Kenya;
4. encourage international travel agents, tour operators and hoteliers to market Kenya as a tourism destination;
5. help private sector organisations to penetrate international tourism markets;
6. co-ordinate tourism marketing projects and programmes, as well as evaluating their effectiveness. (Kenya 1998, 2000; PTLC 1998).

As part of its ongoing management process, the KTB sets targets, mainly expressed in terms of the expected number of tourist arrivals from respective international markets – Africa, America, Asia and Europe. These targets serve as the reference points for planning the KTB's marketing strategies and allocating resources within its marketing programmes (Kenya 1998; PTLC 1998). They also serve as benchmarks for evaluating its overall performance.

However, the initial operation of the KTB was met with leadership wrangles between the Ministry of Tourism and Wildlife (the public sector) and the private sector. The tourism industry's private sector felt that it would manage the KTB more professionally, as it had the 'required understanding of the extreme competitiveness of the tourism market', while the Ministry of Tourism's staff were of the opinion that 'they were more experienced at the task' (source: field notes 2004). This was further compounded by the many state departments in charge of national tourism marketing: the Department of Tourism at the ministry headquarters, which was in charge of tourist offices in key tourist-generating countries; the Marketing Department under the purview of the KWS, and the KTDC and KTB. Fragmentation of the tourism-marketing role between numerous government departments is seen as an obstacle to the construction of an overall national tourism marketing strategy (Lickorish et al. 1991; Hall 1994). The more people involved, the greater the risk that no consensus can be obtained, that views of decisionmakers will differ from one another, or that establishing a community of interests is unrealistic and that departmentalism will prevail (see also Chapter 9) (Hall 1994). Quite consistent with this observation, the creation of the four offices to handle Kenya's tourism marketing resulted in conflicting mandates, and eventually conflicts of interests. This was due to lack of government policy co-ordination, both within and among the four departments. Such co-ordination would help to avoid duplication of roles, and to develop effective tourism marketing strategies.

A near collapse of the national economy in the late 1990s due to unparalleled levels of corruption, cronyism and escalating political violence made the functioning of the KTB extremely difficult. Nevertheless, the KTB relentlessly tried to market Kenya as a destination of choice for 'non-sex' tourists (Kibicho 2004b). This rhetoric aside, a clear relationship between the number of CSWs and tourist numbers (well depicted by tourism seasonality) exists throughout major tourism destination areas (Sindiga 1995; Ndune 1995; Kibicho 2004b). In support of the foregoing, a government tourist officer (name withheld to preserve anonymity) in Malindi 'jokingly' reports:

> The best and probably the most reliable indicator of Malindi's tourism performance is the number of CSWs. The number of sex workers increases as the tourism season improves. In other words, when there are many CSWs in the streets, then there are many tourists in our hotels. By extension, this means that more locals are getting tourist dollars. You know, the trickle-down effect.

Moreover, according to earlier studies on the present topic by this author, this stage of Kenya's tourism development shows undeniable (r)evolution of the local sex industry in the leading destination areas, notably at the coastal region (Kibicho 2004a, 2004b, 2005b, 2006c, 2007). Sex workers and their organisations have become more professional in the way they conduct their sex business (Figure 1.2). Some tourism enterprises, especially hotels, within the formal sector of the tourism industry started to get involved in sex trade-related transactions, albeit in an underground manner (see Chapters 1 and 5) (Kibicho 2005b). In addition, the traditional range of sexual service consumers broadened as female sex tourists flocked to the country seeking romance. Today, Kenya as a (sex) tourism destination is more popular with female (sex) tourists than their male counterparts (see Chapter 5) (Kibicho 2004b). An investigation of factors underlying this popularity would be a fertile area for further research.

Kenya's Tourism Expansion and Sex Trade Growth

In this chapter, the principal aim has been to highlight the value of applying a longitudinal perspective to the influence of Kenya's tourism development on the expansion of sex trade. Based on Butler's (1980) destination life cycle, an institutional approach has been used to illustrate a hypothetical scenario of sex trade growth according to stage of development (see Figure 3.4).

According to Figure 3.4, tourism-oriented sex trade is in its infancy during the exploration stage (the pre-independence period). The trade only involves safari hunters as sexual service consumers. At the involvement stage (1963–78), local inhabitants start getting excited about the prospects of tourism development,

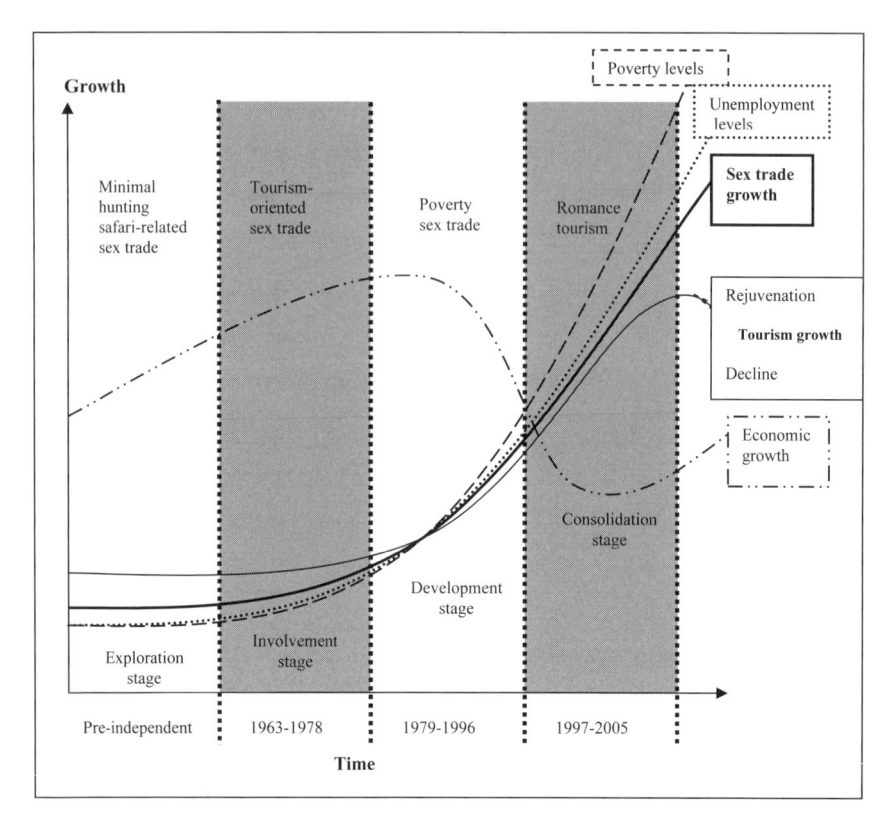

Figure 3.4   Factors influencing Kenya's sex tourism growth – exploration through consolidation phases

commonly associating it with generation of employment opportunities, wealth creation and general individual/community progress. Some local residents start showing an interest in working in the industry. Unfortunately, not all of them can be absorbed into the industry due to lack of required skills (see Chapters 1 and 6). Consequently, a number of them enter into the informal sector of the industry, which includes provision of commercial sexual services. Thus, the involvement stage of Kenya's tourism industry is characterised by an increase in tourism-oriented sex trade (see Figures 1.1, 1.2, 3.4, 4.1, 6.2, 9.1 and 10.1). Exogenous groups, notably scholars and religious leaders, start showing concerns about socio-cultural problems generated by the nascent tourism industry. This occurred in Malindi Area (Figure 6.1) because of the spread of sex trade, the increased number of youths dropping out of school, and heightened levels of alcoholism – all directly or indirectly related to the booming tourism industry (see Chapter 9) (Backwesegha 1982; Peake 1989; Sindiga 1995, 1999; Ndune 1996). However, as happened in Kenya's coastal region, local leaders and entrepreneurs may adopt a campaign to depict tourism critics as 'radical moralists' (Sindiga 1999; Kithaka 2004).

The development stage (1978–96) of Kenya's tourism industry is marked by the establishment of mass tourism (Bachmann 1988; Peake 1989; Dieke 1991; Kenya 1996; Weaver 1998; Akama 1999, 2002; Kibicho 2003). International tourism actors have discovered the destination and are taking keen initiatives in the development of its superstructure and its general promotion. They define, own and sell tourist products. To them, the destination presents an opportunity to make profit (see Chapter 9). Meanwhile, the government is excited about the industry's increased potential to generate even more foreign exchange earnings, and thus maintains a *laissez-faire* business atmosphere. As Figure 3.4 points out, this stage is marked by poor performance of the national economy, characterised by reduced job opportunities and increased poverty levels. This leads to unprecedented rural–urban migrations as many Kenyans leave poor rural areas to try their luck in the urban centres, especially within major tourism destination areas. Due to increased demand for jobs coupled with shrinking employment opportunities, there is an increase in poverty-induced sex trade (see also Kibicho 2004a, 2004b, 2005b). Many local people engage in sex business to generate income for their upkeep as well as that of their relatives (Ndune 1996). Meanwhile, local residents' reaction to tourism development is mixed. Although they begin to recognise certain social problems (such as drug-taking and trafficking, alcoholism), they are also willing to put up with tourism in its current mass-market form because of its associated real and/or perceived benefits.

In the consolidation phase (1997–2005), international tourist arrivals and tourism receipts start to wane. Dissatisfied with the quality of the tourist product, foreign tourists shift to alternative destinations. This causes alarm among policymakers, leading to the establishment of the KTB. However, the KTB is not mandated to transform the existing 'quantity/mass-oriented' product into a 'quality-oriented' one (Kenya 1998, 2000; Kibicho 2003). As a result, Kenya becomes a cheap mass tourism destination. This results in a rise in tourism-oriented sex trade. Sex workers become more professional by establishing informal social welfare associations (see, for example, Chapter 9). This stage is marked by undeniable transformation and diversification of the CSWs' customer base, which gives birth to female 'romance tourism' (Chapter 5). The phase sees an increase in female (sex) tourists visiting Kenya with an interest in consuming (commercial) sexual services (Kibicho 2004a, 2004b, 2005b). In this scenario, if tourism authorities fail to introduce measures to rejuvenate the destination by diversifying the national tourist product while at the same time undertaking aggressive marketing campaigns, the destination will enter the stagnation stage (Figure 3.4) (Butler 1980). The host residents will begin to display increasing hostility towards tourism activities, because of overcrowding, rising crime and the perceived dismantling of local traditions. Fortunately, Kenya's tourism industry does not seem to have reached this stage.

For now, we close this section by noting that due to limitation of space, the model presented in Figure 3.4 only considers a few factors which have direct influences on the growth of sex trade, such as tourism, economy, unemployment levels and poverty. Thus, it leaves out other equally important factors like education, local cultures, legislation, technological advancement and media which might influence the extent to which sex trade expansion dovetails with tourism development at various stages of the destination life cycle. Further, the model is not based on hard data, thus it might not withstand critical scientific scrutiny. Nevertheless, it serves as an important point of departure for those interested in determining how Kenya's sex trade relates to the national tourism industry.

Conclusion

This chapter's major thrust has been a descriptive rather than prescriptive analysis. Nevertheless, based on the foregoing analysis, we are able to draw a number of conclusions. To begin with, since independence, the development of Kenya's tourism industry has been premised on promotion to encourage as large a number of tourists as possible – mass tourism. Currently, the much-vaunted beach and safari tourism are threatened by haphazard development, chronic underfunding and lack of policy co-ordination between various state institutions. The government has failed to design an institutional strategy capable of articulating tourism practices. Such a strategy should enhance local people's participation in the tourism industry in order to remedy the current domination of the industry by transnational capital. It should be noted that increased formalisation of any sector of the (national) economy goes hand in hand with transnationalisation of the industry concerned, which in turn 'destroys' opportunities for small-scale local entrepreneurs. However, the most realistic manner of approaching a future state of sustainable balanced development is not through a single, holistic, enormous step, but by adopting measures that are sensitive to various destination areas' stages in the life cycle.

In closing, it is clear that as a result of the Kenyan government's promotion, tourism has become an important sector of the national economy. This has resulted in continued tourism development, subsequently inducing a massive growth of sex trade, especially within the leading tourism destination areas. This calls for an in-depth examination of the manifestation of the tourism-oriented sex trade in the Kenyan context, which leads us to Chapter 4, 'Sex Tourism in Kenya'.

Chapter 4
Sex Tourism in Kenya

Introduction

As we have seen so far, the issue of sex trade and tourism has been discussed by a number of commentators. Some of the critics claim that 'tourism is prostitution' (Graburn 1979, 1983; De Kadt 1979; Phongpaichit 1981; Cohen 1982; Archavanitkul and Guest 1994; Harrison 1994). This is in line with the metaphorical view that poor countries, as tourism destinations, 'sell themselves' to the rich, tourist-generating countries in order to earn a living. Graburn (1983) sees the poor nations as being forced into the 'female' role of servitude, as being 'penetrated' for money, whereas the outgoing, pleasure-seeking, 'penetrating' tourists of powerful nations are cast in the 'male' role. They are encouraged to open their frontiers and their dwellings to the foreign visitors, and are pressured to engage in commercial transactions of a very particular type in which they offer their culture, their heritage, their traditions and even certain members of their population (Burton 1995). From the locals' viewpoint, tourism is an agent of economic development; but from the outside view, the natives are a mere traditional object of desire.

Taking this as its point of departure, this chapter scrutinises the link between tourism growth and its influence on the expansion of the sex industry in Kenya. Thus, it presents an account of the various dimensions of sex trade in the country which helps us understand how Kenya's tourism facilitates romantic and sexual encounters between local CSWs and (sex) tourists (Figure 1.2).

Sex Trade and Tourism: An African Story

As noted elsewhere, sex is a sensitive subject which respectable students of tourism and other disciplines are reluctant to discuss (Cohen 1982; Thompson and Harred 1992; Sindiga 1995; Ndune 1996; Ryan and Hall 2001; Kithaka 2004). Moreover, until recently, the word 'sex' was a taboo in most African communities (Backwesegha 1982; Sindiga 1995; Ndune 1996; Mungai 1998). Even today, it is uncommon to have the subject discussed freely and openly in these communities (Kibicho 2004b; Kithaka 2004). As a consequence, very few and uninformative studies on the subject of, tourism and sex trade and its different manifestations in Kenya in general have been carried out. In an attempt to explain this apparent lack of academic interest in the subject, Ryan and Hall (2001: i), in the introduction to their book on sex tourism note:

Sex tourism! ... these two words conjure an image of Red Light Districts in places like Amsterdam and Bangkok, or seedy images of men in raincoats boarding buses for massage parlours away from home .... The sex worker is historically a socially condemned person, who challenges norms of the society.

Kibicho (2005b) further notes that any relationship involving sex trade invokes so much indignation and exhortation, thus generating little interest in serious, unbiased and systematic sociological or anthropological research. Consequently, this has led to lack of co-ordinated studies on sex trade in general and how it relates to other sectors of the local economies like tourism in the African contexts. However, Pickering and Wilkins (1993), Sindiga (1995) and Ndune (1996) have done some commendable studies on the subject, though lacking in theoretical, conceptual and methodological coherence – no systematic approach. These works are descriptive, and indeed highly based on secondary data; they rely on data provided by others, sometimes without realising that those authors also derived their data from elsewhere, or used estimates of estimates. Studies by Kibicho (2004a, 2004b, 2005b), however, offer valuable information, and indeed detailed analysis in the tourism context. In fact, the aforementioned three articles are the touchstone of this book.

Link between Sex Trade and the Travel Industry

Sex tourism shares a similar history with that of adventure travel and tourism in general. Literature investigating early travel involving sexual encounters includes stories dating to explorations by Columbus in the fifteenth century (Finnegan 1979). Early journeys abroad that were enhanced by sexual revelries were the 'Grand Tours' – trips taken across Europe by young aristocratic men and women during the eighteenth and nineteenth centuries in order to broaden their understanding of culture and arts (Bauer and McKercher 2003). These adventurous young people often supplemented their cultural experience with that of a sexual nature through liaisons with people they met while travelling (Finnegan 1979; Gallagher and Laquer 1987). Such liaisons involved sexual adventure, either on a non-commercial/casual basis or with a professional CSW (Bauer and McKercher 2003).

When travel opportunities opened up to a growing segment of the middle class in the mid-nineteenth century, sex tourism evolved into a common activity (Caplan 1984). As many European countries became wealthier, clients expanded their search for sex into other regions, such as the Caribbean and northern Africa, where prices for sex were more moderate (Walkowitz 1980). Thus, the expansion of sex tourism has continued unabated, in part as a result of promotion of tourism as a development strategy (particularly in the developing countries) where scarcity of resources and high unemployment levels result in unparalleled levels of poverty. The abject poverty in these countries leads local people, either directly or indirectly, into sex trade (see Chapters 6 and 9) (Pickering and Wilkins

1993). Where tourism has experienced considerable support from the government in resource-scarce regions, for instance, these have proven to be ideal destinations for sex tourists (see Chapter 5) (Naibavu and Schutz 1974).

Sex tourism is expanding worldwide (Naibavu and Schutz 1974; Finnegan 1979; Burton 1995; Cabezas 1999; ECPAT 2002; AVERT.ORG 2004). The principal destinations in Africa are The Gambia, Kenya, Morocco and South Africa (Afrol 2003; Jallow 2004; Kibicho 2004b). As noted earlier, the shift in destinations can partly be attributed to the crackdown in Asia by respective governments (see 'Setting the Scene'). Sex trade has an undeniable multiplier-effect which benefits a range of individuals as well as many economic sectors (see Figure 9.2 and Chapter 9) (Naibavu and Schutz 1974; Symanski 1981; Cabezas 1999; Kibicho 2005b). The first beneficiaries are definitely CSWs, both male and female, themselves and their families. Secondly, hoteliers, restaurant operators, taxi drivers, tour operators and airlines all benefit from people who travel to participate in one way or another in sexual activities while away from their usual places of residence. Overall, these sex workers are in no small degree responsible for tourist spending while in the tourism destination, thus contributing to much-sought after foreign currency (Figure 9.2). Unfortunately, due to the illegal status of the trade in many countries, accurate statistics about CSWs, their clients and the money generated by the trade are unavailable.

In general terms, the expansion of sex tourism has continued, in part, as a result of the promotion of tourism as a key development strategy, especially in the less developed countries (Graburn 1983; Ryan and Kinder 1996; Seabrook 1996; Cabezas 1999). Of course, tourism only provides an exit route to poverty-related predicaments. Subsequently, tourism-related sex trade is today a multi-billion-dollar industry that supports a major portion of the international labour force (Archavanitkul and Guest 1994; Cabezas 1999; Ryan and Hall 2001).

Kenya: A Cheap Haven for Sex Tourists

In Kenya, street CSWs open negotiations by asking for a higher amount of money, for example US$70, for oral sex (see Table 9.3 for a detailed price list of sexual services). This leaves room for negotiation. The quoted amount can be reduced by more than half if the potential client is a good negotiator. In some instances, inexperienced CSWs can be persuaded or 'tricked' into spending a whole night with a client for the cost of a meal or a few drinks worth less than US$5 (see also Chapters 6, 8 and 9). In relation to the foregoing, a German (sex) tourist in Mombasa (Figure 6.1) reports:

> it costs less to spend days, weeks or even months indulging oneself in Kenya than in other countries, such as the Philippines, Thailand and Vietnam. This is probably because gals [CSWs] here are independent and thus they are not paying commissions to pimps .... Stiff competition between these gals [CSWs]

also lowers prices. This favours us [sex clients], who are treated like kings … and of course, we are kings here … kings in the Kenyan bedrooms. … Aren't we? [he asks sarcastically]. Prostitutes will entice tourists away from each other with offers of better deals – cheap accommodation plus sexual access. They occasionally fight over me [he concludes boastfully].

However, not all sex tourists in Kenya prefer multiple and anonymous sexual encounters. Some, it seems, can only attain sexual and psychological satisfaction from a CSW after they have convinced themselves that they are in a reciprocal relationship. Consequently, such sex tourists prefer to spend their holidays with the same CSWs. They do not consider themselves as 'clients', and do not consider their partners as CSWs, but rather as companions (see also Chapters 6 and 9). They often turn down CSWs who approach them with direct sexual propositions. They prefer less explicit overtures, such as:

Where are you from?

Do you like Kenya?

Do you like Mombasa/Malindi/Mtwapa?

Where do you stay?

Do you like your hotel?

Are you alone on this visit?

You like Kenyans?

Kenyan women/men [euphemism for sex] are good, eh …?

This gives the potential sex tourist an opportunity to continue as in non-commercial encounters:

What is your name?

Can I get you a drink?

Would you like a dance with me?

Would you like to have dinner with me?

Would you like to see where I am staying?

The whole process can then be interpreted as confirming a mutual attraction. Consequently, when the CSW later confides his or her desperate financial needs, the sex tourist can construct the act of giving money not as payment for sexual services rendered, but as a gesture of appreciation and helping a friend in need (see also Chapters 6, 8 and 9). However, in some instances, due to the fact that CSW–tourist encounters are 'rare', sex workers are 'forced' to be rather explicit about their intentions. This is mainly because these young Kenyans work under pressing limits of time and fierce competition from their peers (see also Chapters 8 and 9).

As well as granting 'sexual license', a female CSW often helps the tourist to find cheaper accommodation, sometimes putting him up in her own room, acting as his guide and interpreter. She may even do his laundry and cook for him. In return, he is expected to pay for food, drinks and evening entertainment. He may buy for her soap, shampoo and clothing, or leave her some cash when he moves on to the next destination or to the next 'girl'. The price paid by the sex tourist and the benefits secured by the CSW are thus variable (see Chapter 9). An experienced CSW could easily manage to squeeze as much as US$450 a day from a client. However, not all of this will be in cash. On the other end, as noted above, an inexperienced CSW may end up offering sexual services for almost nothing. To qualify this, Jensen (not his real name), a Dutch (sex) tourist, reports that 'some of them have slept with me for just a bar of soap costing less than one US dollar'.

The sums of money involved are often negligible to sex tourists from the developed countries (see Chapter 5). Karl (pseudonym), a (sex) tourist from Berlin, for instance, notes:

> my 'girlfriend' suggested that I move out of the hotel where I was paying US$74 per night, and stay in her apartment. Here she does all my washing and prepares meals for me. For all this, plus acting as my local guide and ... of course granting me sexual access, she asks from me only US$25 a day plus the cost of food ....

Although this tourist was 56 years in age, fat with receding hair, while his 'girlfriend' was 18 years old and, in his words, 'a potential miss Africa – very beautiful', he felt that this money was too little to have anything to do with the invitation she had extended to him: 'She must have found me sexually attractive to be offering so much for so little in return,' he concluded with a contagious laugh.

Contrary to this tourist's conviction, this is a common strategy of a female CSW trying to appeal to the Western man's perception of himself as a 'real man' who can take care of his 'woman' – of course, with the ulterior motive of getting more pay for it. Thus, the relationship is no longer perceived as a CSW–client liaison, but a relationship between two consenting adults, thereby legitimising the whole affair. In simple terms, such 'stage-managed' behaviour by a CSW is a fundamental strategy to prolong the relationship, and in turn, to extend its profitability. A CSW who is able to attain this goal – prolongation of a CSW–tourist relationship – is looked upon by colleagues with respect and envy (Cohen 1993; Günther 1998; Oppermann 1998; Phillip and Dann 1998; Kibicho 2005b).

Moreover, from an economic frame, such suffering of some loss by the CSW can be accounted for as an investment cost, which will be overcompensated by future 'profits'. In other words, sexual contacts and renunciation of payment by a CSW can be well-calculated moves aimed at encouraging a lasting emotional and, as a result, economic commitment by the tourist.

The remainder of this chapter examines sex tourism in Kenya from three global perspectives: (1) the main types of sex trade practised in Kenya; (2) the principal types of commercial sexual services offered in the country, and (3) advertisement of commercial sexual services. However, it is necessary to note that these services are not assembled solely for foreign tourists. Sex workers also target the local market, which in most cases consumes a higher percentage of sexual services. Nevertheless, as noted in Chapter 1, it is not the intention of this text to examine tourism-related and non-tourism-oriented sex trade separately. Domestic and international tourism-related forms of sex trade are also evaluated together.

Types of Tourism-oriented Sex Trade in Kenya

Sex trade in Kenya takes different forms (Kibicho 2004a, 2004b). These include brothel-based sex trade, street sex trade, striptease and escort or call-out sex trade. The remainder of this section discusses these four types of sex trade.

Brothel-based and Private Apartment-based Sex Trade

Brothels are establishments specifically dedicated to sex trade, often confined to RLDs in major urban centres (see Chapter 1). In Kenya, a brothel is also known as *nyumba moto* ('hot house'), house of accommodation, meat house, public house or massage parlour. Of importance to note is the fact that massage parlours serve as fronts for brothels in the country.

Sex trade can also take place in a CSW's apartment, and in many countries this is the only 'accepted' form of prostitution (Archavanitkul and Guest 1994). A hybrid between brothel and private apartment prostitution exists in Kenya – CSWs rent one-room apartments in RLDs then solicit clients, either directly or through pimps (see Chapter 8). This form of sex trade is common in Majengo and California estates in Nairobi, Kizingo Estate in Mombasa, Salgaa Centre in Nakuru, Huruma Estate in Eldoret, and Kisumu Ndogo and Embassy estates in Malindi. From a general point of view, brothel-based and private apartment-based sex trade in Kenya accounts for about 25 per cent of the national sex business (Figure 4.1). Less than 10 per cent of the CSWs in this category of sex trade are within the tourism sex trade, while the remainder (over 90 per cent) lie within the lower end of the three-tier sex trade triangle – poverty sex trade (Figure 1.1).

However, there are no bonded CSWs in Kenya. Bonded CSWs are sold to brothels in order to repay debts or reduce loans. According to McLeod (1982) and Brown (2000), bonded sex trade is a form of slavery in which the body of the CSW is treated as a commodity to be bought and sold without the sex worker's consent. Further, some authors argue that this form of prostitution takes sex from being a soul–body relationship to a subject–object one (Hall 1996; Afrol 2003; Agrusa 2003). According to Graburn (1983) and Ackermann and Filter (1994), the subject is the man who buys and the object is the woman who is sold. None the less, the discussion on the merits and demerits of bonded sex trade is beyond the scope of the present text.

Street Sex Trade

Street sex trade is the most prevalent and most conspicuous form of prostitution in Kenya (Mwakisha 1995; Kibicho 2004b; Kithaka 2004). Overall, street-based sex business makes up about 60 per cent of Kenya's sex trade (Figure 4.1). Almost 94 per cent of the CSWs in this group are within the poverty sex trade (Figure 1.1). In this kind of sex trade, a CSW solicits customers while waiting at street corners or walking on a street. They are usually dressed in skimpy and suggestive clothing. Often the sex worker waits for a potential client to initiate the process. However, Kenya's street-based CSWs have perfected both implicit and explicit ways of making their presence noticeable even to those not interested in their services.

The agreed-upon sexual service is normally delivered in a rented room, in the customer's car or in a nearby alley. This type of sex trade offers the CSW a high level of freedom. Sex workers are able to set their own prices. They have the freedom to choose their dates, as opposed to receiving clients they have not planned for in a hotel room, as is the case with brothel-based sex trade. However, it can also be the most dangerous form of sex trade, as CSWs sometimes fall into the hands of dangerous clients (see, for example, Chapter 7). Of importance to note is the fact that, many 'brothel' workers and call girls in Kenya eventually end up as street sex traders. There are several reasons for their return to the streets:

1. they may not want to continue paying commission to brothel owners and/or pimps;
2. they may have extreme drug addiction problems such that brothels nor sexual service agencies are no longer willing to transact sex business with them;
3. they may not have saved enough money to start an alternative business in their old age, when CSWs are less attractive to clients and thus become a liability to brothel owners;
4. they may have been disfigured through beatings by clients or by brothel owners or pimps, thereby becoming sexually unappealing to clients.

Striptease

A striptease is a performance, usually a dance, in which the performer gradually removes clothing for the purposes of sexually arousing the audience. Teasing involves the slowness of undressing, which makes the audience eager to see more nudity. The act of undressing is accompanied by sexually provocative movements. Striptease performers are called strippers or exotic dancers.

Striptease has come a long way since the breathtaking presentation by Adah Menken – who, in 1861, tied to the back of a horse and clad in pink tights, sent her 'spectators' beyond the hissing lights into paroxysms of enthusiasm (see Bauer and McKercher 2003). There is little to differentiate between contemporary striptease and that of the past. However, modern fancy costumes complemented by sophisticated computer-controlled lighting systems help to draw the clients and the performer closer emotionally, albeit for a brief moment. There are both female and male strippers in Kenya, but female striptease is more common than male striptease in the country. Generally, striptease sex trade accounts for about 9 per cent of Kenya's sex business (see Figure 4.1). These strippers mainly serve the tourism industry, and thus are within the tourism sex trade (Figure 1.1).

Female strippers A Kenyan female stripper entertains her clients, in most cases men, with seductive movements timed to the rhythms of music. Typically, she performs to a set of three songs. Ordinarily, the performer has to wait until the third song before she starts to remove her costume, eventually revealing her breasts and the 'T-bar'. In most cases, the performance ends as soon as the undressing is completed. The end of the third song marks the beginning of the first song for the next exotic dancer. This procedure is repeated throughout the evening.

Lap dancing, contact dancing and friction dancing are advanced forms of striptease. 'Floor work', normally the most erotic part of the stripper's stage repertoire, allows an individual exhibition of prowess as well as a generic display of sexuality. Normally, this is done during the last song of a performer's set. In addition to stripping, the performers also offer 'private dances', which involve more attention for individual members of the audience. The individual is chosen from the audience by the performer, mainly based on the level of interest he shows during the performance. This level of interest is determined by the amount of money the individual pays in the form of a tip during the dance. In fact, some performers use these 'tipping' signs to identify customers willing to exchange cash for a brief 'visual offering' – false intimacy. These events are thus play for the clients, but work for the stripper.

The contact can vary from a simple 'up-close dance' with no touching to physical contact with the stripper, or even sexual intercourse, though this is rare in Kenyan clubs. Some clubs have booths where customers can take a shower with the strippers for a fee. Others have pits with pudding, oil or mud wrestling, allowing for closer contact between the patrons and several nude performers.

Variations on this theme of female strippers include table dancing, where the performer dances on or by the customer's table, couch dancing, where the client sits on a couch, and sexual acts between two female strippers, such as touching, kissing, fingering and cunnilingus. In Kenya, these specialty acts go by such names as 'Big Titis', 'Bobbie Balloons', 'Mrembo Boobies', 'Manyanga Boobs', 'Lady Curves', 'Rocki Mountains' and 'Letha Weapons'.

According to Zainabu (not her real name), a female stripper in Mombasa, women are lured into striptease by the huge tips they receive during the performances. She reports:

> Normally, I am paid K£10 for each hour I work in the club. On a good night, I get tips (both in cash and in kind) equivalent to K£700. I work here because of the tips and not the salary. Generally, we [CSWs] get good tips from cruise tourists. They also pay well for other sexual services ....[1]

Male strippers Until the 1970s, strippers were almost invariably female, performing to male audiences (Jago 2003). Since then, male strippers, performing to both male and female audiences, have also become popular. It is important to note that male and female strippers also perform for gay and lesbian audiences respectively. Before the 1970s, however, male striptease performances appeared largely in 'underground clubs', but the practice has eventually became common in many sex tourism destinations (Bauer and McKercher 2003; Jago 2003). These kinds of clubs are also springing up in Kenya (Kibicho 2004b, 2005b). For example, the author visited two such clubs in Nairobi, one in Mombasa and one in Malindi, during the study period. However, all of them are very secretive and selective about admissions being only for registered members. Despite being a non-member, this author was admitted to these clubs courtesy of the proprietor of the striptease club in Malindi. Nevertheless, the author had to promise that all the data gathered would be handled with the utmost confidentiality and that the clubs' anonymity would be safeguarded at all times.

It merits mentioning that although male striptease is currently not very common in Kenya, it is just a matter of time before it becomes a popular feature of the country's sex industry. This observation is based on the high levels of enthusiasm exhibited by the clients in the four clubs visited during this study. Rashid (fictitious name), a proprietor of one of the strip clubs notes:

> people are looking for something different to do, something with a tactile element. In the past, Kenyans would go out, talk and smoke, drink and eat, sing and dance. I think that is gotten kind of old fashioned. Now, people want a human touch in their outings .... That is exactly what we are offering in our clubs. And, Kenyans appreciate our efforts by patronising our business in large numbers.

Unfortunately, the club proprietors complained of the huge amount of money they are paying to the 'security operatives'. They call it a 'protection fee'. This illegal fee, which is corruption *par excellence*, is paid to some senior security personnel

---

1 See Chapter 6.

in the government on a monthly basis. In return, senior security officers ensure that club owners and their clients are not 'harassed' by junior law enforcers.

Visits by women to clubs featuring male exotic dancers are common in Kenya. Unlike the enforced sedate atmosphere at clubs featuring female exotic dancers for male audiences, the female audience for male strippers is sometimes very aggressive. Some female patrons, for instance, join male exotic dancers on the stage, assist the performers to strip, and in some instances strip together with them. This is possible because male exotic dancers do not view women clients as physical threats, as female dancers would if their male patrons were to join them on the stage. As a result, the nightclub management and their security men (bouncers) do not restrain their female audiences from interacting closely with the male strippers. Nevertheless, the bouncers do not allow female patrons to have sexual intercourse with the performers during the presentation. This can only take place after the show. The club management must be informed of such arrangement, however, as they must be paid a percentage of the total fee.

Conversely, gay strip clubs (featuring male strippers for a male audience) rarely feature male performers who do a complete striptease. In general, they feature men who appear initially in skimpy and sexually revealing undergarments. Unlike in female striptease, though technically prohibited, fondling the strippers is common. In cities such as Amsterdam and Paris where full nudity is allowed, the male strippers at gay venues masturbate on the stage to maintain an erection. Sometimes they allow willing members of the audience to participate in this masturbation process. However, one has to pay for such participation. The payment is normally in the form of a tip. Many gay strippers prefer this mode of payment, as they are not required to pay a commission to the hosting club. Based on discussions with the proprietors of the four striptease clubs visited, however, gay strip clubs do not exist in Kenya.

From a general viewpoint, striptease performances allow human interaction to take place in a special commercial setting which shields both strippers and clients from the ordinary effects of transgression. The process becomes one of socialisation, temporarily dissolving social barriers to allow for role malleability, as participants are free to act in ways, and to act out roles, perhaps unacceptable in other (public) places (Bauer and McKercher 2003; Jago 2003).

Although the emotions displayed in both male and female performances might be authentic, they are derived from fictive rather than factual stimulus. However, it is worth remembering that authenticity is a subjective phenomenon, the meaning of which depends upon the subjects, their expectations and willingness to make-believe. Nevertheless, it is a dynamic, emergent phenomenon, implying that today's staged phenomena may turn into authentic ones tomorrow (Cohen 1998a). Moreover, we tend to see what we expect to see, and overlook that which is not in conformity with our pattern/expectations. This allows for a creative invention and presentation of deeply imagined human qualities, giving birth to an ambience conducive for relaxation, socialisation and excitement. Indeed, some patrons are temporarily lost in fantasy, whereas others are recharged from personalised attention, and/or are excited by the escape from work and family obligations. Consequently, it may be argued that the specificity of the journey to the striptease clubs constitutes a kind of micro-touristic event. Obfuscated within this kind of discussion, however, is the extent to which authenticity is created and developed. It is definitely a delicate balance between 'complementary commercialisation' and 'substitutive commercialisation'. In other words, these performances can be produced *en masse* in order to satisfy the mass of tourists.

Escort or Call-out Sex Trade

In escort or call-out sex trade, clients calls an agency and the sexual service delivery takes place at the customers' place of residence, or more commonly at their hotel room (Harrison 1994; Brown 2000; Bauer and McKercher 2003; Jago 2003). The escort agencies also supply attractive escorts for social occasions (Kibicho 2005b). While many escort agencies provide sexual services (see Chapter 1), some ignore escorts who provide additional sexual services.

An interested client contacts the agency by telephone and offers a description of the kind of escort they are looking for. The agency then proposes a CSW who might fit the client's requirements. The agency then informs the selected escort, who in turn contacts the client to arrange for an appointment. It is during this first contact that details of the sexual services requirements are discussed. Then the escort calls the contracting agency upon arrival at the agreed location, and again upon leaving, to ensure safety.

Alternatively, an escort may operate independently of an agency. For example, most of Kenya's escorts and call-outs are not attached to escort agencies as in the Western countries. This form of sex trade comprises 6 per cent of the national sex business (see Figure 4.1). Sex workers in this category exclusively serve tourists and local elites. They are all in the high-class sex trade (Figure 1.1). Sex workers who operate as escorts normally advertise their services in local newspapers and/ or magazines. As we shall see later, these advertisements carry sex-suggestive names which focus more on the sexual offerings to be expected. Names like 'sexqueens', 'playboy', 'maletoys', 'dreamlands', 'Kenyancunts', 'pussygalore' and 'privaterooms' need no introduction to CSWs' clients as they leave little doubt as to the business's real nature. The connotative value of words can often give dual meanings to phrases that are written on services promotion materials, and in most cases these vague meanings could be used to refer to commercial sexual services.

In Kenya, there is a growing trend for university and college female students to work as call girls to pay for their college fees and other personal expenses. Working as call girls is compatible with these girls' routines as they only work during the evenings and at weekends. This enables them to attend classes during the day. An example is Judy (not her real name), a fourth-year student at a university in Nairobi. She is one of the most sought-after call girls by night, and plies her trade enthusiastically at the weekend. Her sexual services have price tag of US$120 per hour or US$230 for a whole night. Judy's earnings per night put to shame the paltry amount street CSWs earn in Kenya's coastal region and other urban centres (see, for example, Table 9.3). With this kind of money, Judy can afford to rent a three-bedroomed apartment in the more affluent suburbs of Nairobi, where she entertains the medley of the rich and famous who feature in her list of clients. The list reads like a 'who's who' of business circles, the political arena and government sectors in the city of Nairobi. She lives in an equally big house in the same neighbourhood.

Meeting Judy was not easy, however. I had been given her cell phone number by one of her friends who worked as a CSW in Malindi. I pretended that I was a potential client. When she realised I was only interested in her story, she abruptly hung up on me. She called me after two weeks to accept my request for an interview. When I asked her why she had changed her mind about the interview, she said:

> At first, I thought you wanted to set me up ... you know, sting operation. I have suffered before .... I thought about your request, and decided to talk about it because I like trying out new things and I am not ashamed to admit that some of the things touching on sex are out of the ordinary.

However, she set two conditions for this interview – that her identity be kept secret, and that the researcher should not judge her for who or what she was or what she did.

On her earnings and future plans, she reported:

> On average, I make about US$3,200 a month. My clients are mainly married businesspersons, civil servants and politicians who are looking for fun. I get over 80 per cent of my clients from referrals. I also advertise my services .... Because this job is stupidly categorised as illegal, I only put my cell phone number in the local newspapers followed by eye-catching words like 'male fantasy parlour', 'Kenyan hot pussy-kat'. I will continue with this work after my university studies. I hope this important business [sex trade] will have been legalised in Kenya by then. I also intend to write a book about my experience.

Most of these 'girls' quit this profession upon completion of their studies. Others continue to work in the trade until either they get married (though some continue even after getting married) or they become 'too old' and thus can hardly attract 'well-paying' clients.

From a general viewpoint, Kenya's street sex trade, brothel-based prostitution and private apartment-based sex business lie at the lower end (poverty sex trade) of the three-tier classification of sex trade (Figure 1.1). Striptease falls under tourism sex trade, as the CSWs' cash retention is relatively higher. Access to this kind of sex trade is also not open to all, thus there are fewer CSWs in this group. The number of sex workers decreases further when it comes to the escort or call-out sex trade, while the cash retention levels increase. This goes hand in hand with improved professionalism. Thus, this category lies within the top of the sex trade

Sex Tourism in Kenya

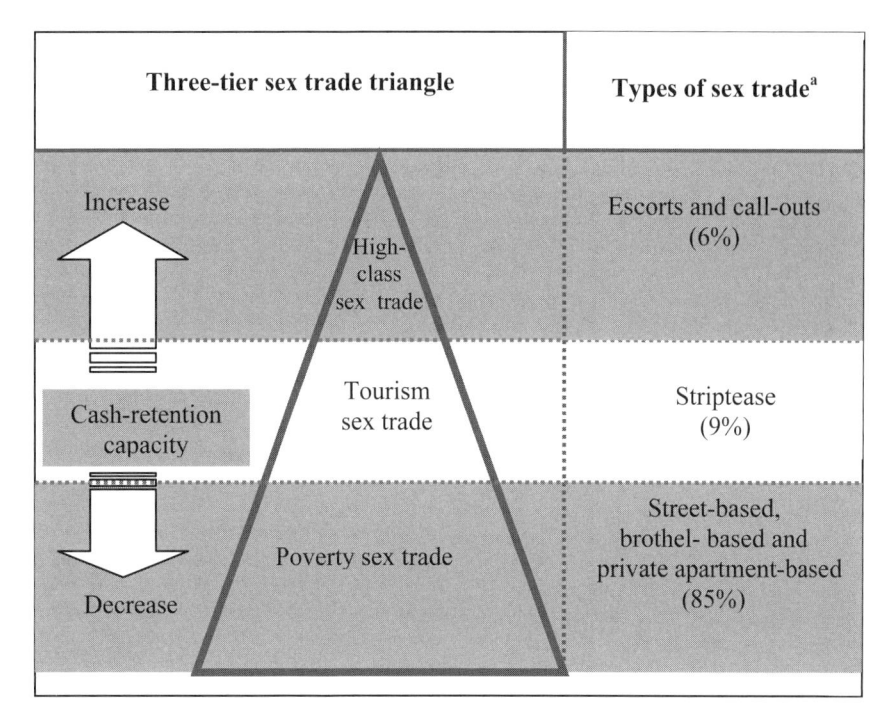

Three-tier sex trade triangle	Types of sex trade[a]
Increase ↑ Cash-retention capacity ↓ Decrease — High-class sex trade / Tourism sex trade / Poverty sex trade	Escorts and call-outs (6%) / Striptease (9%) / Street-based, brothel-based and private apartment-based (85%)

Figure 4.1   Principal types of sex trade in Kenya

Note: [a]Figures in parentheses are percentages in relation to national sex business.

spectrum as detailed in Figure 1.1. Consequently, Kenya's tourism-oriented sex trade can be presented as in Figure 4.1.

According to Figure 4.1, 85 per cent of Kenya's sex business can be categorised as poverty sex trade. It comprises street sex trade (60 per cent) and a hybrid between brothel-based and private apartment-based prostitution (25 per cent). Subsequently, a minority of CSWs in the country – 15 per cent – offer either striptease sexual services (9 per cent) or escort and call-out sexual services (6 per cent).

Principal Types of Commercial Sexual Services in Kenya

As noted in Chapter 2, commercial sexual services take place any time and anywhere (see also Chapter 6). However, there are some venues which are more popular than others. In Kenya, for instance, the main places for selling sexual services are bars, massage parlours, nightclubs and discothèques. Commercial sexual services offered in these venues are generally packaged in two forms: body massage and penetrative sexual services.

Body Massage

Often, body massage in Kenya consists of an oil massage and then a bath. In the sex trade circles, a combination of these two services is referred to as 'body-slide treatment' or 'soapie'. In most cases, this is followed by sexual services. Normally, the choice of the type of commercial sexual services to be offered is left to the discretion of the client. The charges for the sexual services are sometimes not included in the price paid to the hosting club. If it is not negotiated together with the rest of the services, it is treated as a side service by the masseuse, and will thus be paid for separately.

Penetrative Sexual Services

Express penetrative sexual services in Kenya can be purchased mainly in bars, body massage parlours and discothèques. Some discothèques are similar to strip clubs in Western countries. However, dancers in many of these clubs are not supposed to be fully naked. Instead, they dance topless or in a bikini, or in similar revealing costumes.

Sex workers, barmaids, are employed by bars either as hostesses in the hostess bars or as dancers, in the case of discothèques. Both the hostesses and the dancers are also required to market their sexual services, mainly through personal selling or one-to-one publicity. In most cases, clubs employ one or more *(wa)mamas* –relatively old and experienced female CSWs – who match interested customers with their preferred CSW. A customer pays a flat fee to the hosting club in order to leave with his selected CSW. However, the payment for the actual sexual service is not finalised at this stage. The client will have to negotiate for this service with the CSW concerned. These sexual service charges generally depend on the amount of time the CSW will spend with the customer. These services are divided into two:

1. short-time – at most, a duration of three hours; this time can vary from one club to another, and is also referred to as the 'grace period' – a time when the CSW is not charged for being with a client;
2. long-time, or overnight.

The short-time sexual service is guided by the fact that the CSW is supposed to return to her hosting club after three hours. At the expiry of this time, the CSW pays a commission for every extra three hours spent outside the hosting premises. This commission is calculated using an agreed rate for short-time sexual services by the club in question. Consequently, if both parties (the CSW and her client) decide to prolong the relationship outside bar hours, the onus then lies with the sex worker to make the relationship profitable (for her), while at the same time negating the conventional perception of sex trade – money for sex, exchange value for use value, perception.

Some bars have rooms from where CSWs offer short-time sexual services – short-time rooms. In Kenya, this casual in-room sexual service is called *pillow moto*, or 'hot-pillow trade'. Services such as provision of clubrooms for hot-pillow business have a number of characteristics that distinguish them from industrial products. They include intangibility, inseparability, heterogeneity and perishability (see, for example, Kibicho 2005c). The distinguishing characteristic that influences marketing strategies in the hot-pillow trade and in the service industry in general is perishability. The fact that services are perishable means that they cannot be stored to be sold later. An unpurchased club room is an opportunity lost for ever. However, while it is popular to talk about the perishability of the (sexual) services, the often-forgotten flip side of that characteristic is that the service/product can be sold over and over again. Doing this, of course, depends upon demand. Through hot-pillow trade, for example, it is possible for the club operators to occasionally clean and then re-rent their rooms after the clients have left. This leads to an occupancy rate of more than 100 per cent.

Many discothèques in Kenya do not employ permanent CSWs. Instead, they allow 'freelance' sex traders to solicit clients within their premises. Most tourist-oriented nightclubs also allow for similar solicitation. In such instances, the differences between bars and brothels in Kenya in general and in the coastal region in particular seem to have blurred – they offer similar commercial sexual services. In one bar in Malindi, for instance, female employees offer oral sexual services to clients at the table, to enable men clients to continue to chat and drink with their friends during the (sexual) service delivery process (see Kibicho 2004a).

Advertising Commercial Sexual Services in Kenya

For simplicity, the term 'advertising' is used here to refer broadly to all forms of persuasive marketing efforts whereby the marketer wishes to communicate a given sales message to a specific customer audience regardless of the medium used. Such activities include point-of-purchase promotions, as well as print, direct mail, and broadcast advertising.

In countries where sex trade is legal, advertising for commercial sexual services might be legal or illegal. In the Netherlands, for example, it is legal to advertise these services, while it is illegal to do the same in Germany despite the fact that sex trade itself is legal. As we noted in Chapter 2, sex trade is illegal in Kenya. Likewise, it is also illegal to advertise its related services. However, advertising of commercial sexual services does take place in the country, albeit in clandestine forms, ranging from cards in newsagents' windows or stickers on street light posts to messages in the World Wide Web.

Cards in Newsagents' Windows or Stickers on Street Light Posts

In this kind of advertisement, the CSW gives a name followed by a telephone number. In all the cases, the name is fictitious but the telephone number is real. According to Stella (not her real name), a female CSW in Malindi (Figure 6.1):

> by giving fake names, we [CSWs] hide from those who know us. In fact, the telephone numbers we [CSW] give are purely for business. For example, none of my family members or friends knows my cell phone number .... It saves us from being judged ... and being stigmatised.[2]

Regular Magazines and Newspapers

Commercial sexual services marketing through regular print media (magazines and newspapers) takes the form of euphemistic advertisements disguised as 'massage' and/or 'relaxation' services. This is intentional, in order to pacify public opinion. Of course, few will be fooled into believing that, for instance, massage parlours in Kenya are anything different from brothels. Yet, by some twisted social logic, the usage of this 'public-friendly tag' (massage parlour) ensures their existence. In Kenya's coastal region, for instance, massage parlours are quasi-legal fronts for sex trade, because massage is legal, but sexual activities which take place in these parlours are illegal.

Tart Cards

Tart cards are less than half-page documents with contact information – a name (normally a pseudonym) and a telephone number. As in regular magazines, tart cards carry euphemistic information only discernible by regular users of the commercial sexual services. They are normally placed in public telephone booths. Some CSWs also use bumper stickers to advertise their sexual services. However, the major challenge in this type of advertisement is to convince car owners to fix the stickers on their cars. Diana (fictitious name), a female CSW in Mtwapa (Figure 6.1) explains:

> Many car owners do not like to associate with us [CSWs], especially in public. As a result, only a few of them accept to carry my stickers on their cars. However, I have three beautiful ones [stickers], two on the sides and one at the rear of my Toyota RAV4. I had them made by an IT professional in Mombasa.

---

2 See, for example, Chapter 6.

Specialist Contact Magazines

Sex workers in Kenya also use specialist contact magazines. They include *Life Seen*, *Love Contacts*, *True Love* and *Hotpepper*. Unlike in the regular magazines, advertisements in these local publications are explicit. Laurie (fictitious name), a CSW, noted:

> there is nothing to hide here ... a rose by any other name would smell as sweet. Similarly, a pussy cannot be a pussycat ... [she smiles]. If you buy these magazines, you already know what you are getting into. It is called being responsible .... Consequently, we [CSWs] do not mind having our nude photos in these magazines. In any case, as for me, I am proud of my beautiful body. Don't you like my legs? And my boobs [she tickles the author].

The World Wide Web

The World Wide Web is another crucial avenue for marketing commercial sexual services in Kenya (see, for example, Chapter 5). However, only a negligible number of CSWs in Kenya have embraced this form of advertisement. Even the elites of the sex business – those in high-class sex trade (Figures 1.1 and 4.1) – with their relatively high levels of education and professionalism seem to be unaware of the immense power of information technology in the field of sex/tourism services marketing.

Conclusion

Sex trade has for a long time been a means by which governments of a range of developing countries have obtained increasing amounts of foreign income from international tourists. In some cases, however, official government data on CSWs are lower than the actual figures, probably because they are an embarrassing reflection on government ideology, and indeed policies. In Kenya, for instance, tourism growth has brought about proliferation of sex trade in some tourism destination areas. However, until the release of UNICEF's report on child sex tourism in 2006, the government had been in a constant state of denial of the fact that sex tourism is a major component of the national tourism industry. Unfortunately, this campaign of denial is self-deceptive, self-destructive and ultimately self-defeating. As a consequence, for example, the theoretical framework for analysis of the relationship between sex trade and tourism in the country remains fragmentary and contradictory. This necessitates studies into the nature of sexual politics and the inter-related structures of gender relations in order to reveal more insights into such relationships. This thesis forms the heart of Chapter 5, 'Kenya's Tourism Industry: A Facilitator of Romance and Sex'.

Chapter 5
Kenya's Tourism Industry: A Facilitator of Romance and Sex

Introduction

Cheap airfares, the 'opening up' of countries once 'closed' because of war or political crisis and the advent of the Internet have provided more opportunities for tourists looking for short-term sex partners. Countries such as Cambodia and Vietnam have been well-known destinations for such tourists, but now Thailand, the Philippines and Sri Lanka have gained similar reputations (Mingmong 1981; Burton 1995). This scenario is also spreading to other developing countries in Africa, in Eastern Europe and in Latin America (AVERT.ORG 2004). Today, Kenya is undeniably one of the key sex tourism destinations in Africa (Sindiga 1999; Kibicho 2004a). The country has become a favourite destination for sex tourists, mainly due to lack of policy guidelines to control its ever-expanding sex industry (see Chapter 3).

This chapter examines how Kenya's tourism industry creates a market for commercial sexual services. It also discusses the principal enablers for the expansion of tourism-oriented sex trade in the country. Thus, the chapter continues the discussion in Chapter 4, while at the same time laying the foundation for a comprehensive scrutiny of the manifestation of sex trade in Kenya's coastal region – the bedrock of the country's tourism industry (see Chapter 6).

Tourism as a Creator of Sex Trade

Many African governments have adopted tourism as a strategy for development (De Kadt 1979; Sinclair 1990; Dieke 1991; Hall 1994). This is partly due to the incessantly declining terms of trade and the pressure on these governments to diversify their economies from the all-powerful international lending agencies, notably the World Bank and the IMF. Scarcity of much-needed foreign exchange further serves as a strong justification of the adoption of the tourism industry as a principal generator of hard currency. In consequence, despite tourism's adverse long-term socio-environmental effects, its economic benefits, which are sometimes overestimated, seem to override all policy and planning decisions (see, for example, Chapter 3).

Macro-evaluations of the tourism industry usually stress the economic gains it has brought to a given destination, without considering the socio-cultural negative effects – disruption of the value systems and life patterns (Burns and Holden 1995; Brown and Hall 2000). Socially, the problems posed by tourism are complex (Seabrook 1996; Alleyne and Boxill 2003). In Kenya, for instance, tourism resorts have sprung up in areas where local people have been living in abject poverty. Luxury hotels are suddenly opened in their midst, with consumer articles they have never seen before. They see foreign people sunbathing (while half-naked) and/or strolling in the shopping centres (purpose-built for them) displaying expensive trinkets as a show of their financial might (see, for example, De Kadt 1979). It is natural, therefore, that the poor local residents are tempted to take part in the tourism industry and make some money out of it. Sooner rather than later, they realise that formal employment opportunities are limited. Faced with these realities, they seek other means to profit from the industry, including illegal ones (Kibicho 2004b). The unceasingly expanding sex trade, both female- and male-oriented, not to mention child sex tourism, is clear proof that other means of livelihood within the mainstream tourism industry are unavailable to them or are not as profitable (Peake 1989; Kibicho 2004b).

The tourism industry, like other modern industries within a capitalist set-up, is based on profit-making and exploitation. In other words, the business of tourism is to make money (see Chapters 3 and 9). This is often expressed by describing business as self-interest mediated by conditions of the market (Dahles and Bras 1999; Jago 2003). In line with the foregoing, tourism in Kenya flourishes by exploiting exotic natural resources for profit (see Chapter 3). For tourists' entertainment, the industry depends on a series of unequal relations between the powerful and the powerless, the wealthy and the poor – the visitors and the visited. At times, these skewed tourism–host community relations are insensitive to the value systems of the local residents (see Figure 9.2; Chapter 9). Such insensitivity may lead to social havoc and irrevocable damage to people who are ordinarily ill-equipped to handle such abrupt changes in their lives.

The tourism industry can also be seen as a facilitator of romance and sex. This can be envisaged on a continuum that ranges from simply providing the setting to providing specific infrastructure such as accommodation facilities, nightclubs, massage parlours or even brothels. Further, this facilitation can range from simply bringing people together to undertake a tour such as a safari to PAs, to prearranging sex tours to foreign destinations. As mentioned in Chapter 2, some Japanese tour operators, for example, have been known to organise sex tours. The Japanese sex tours to South East Asia peaked with the 1964 liberalisation of travel abroad and the growth of the country's economy. One subsidiary of a car manufacturing company was reported to provide overseas trips for its most productive workers to countries in the South East Asia region. Tomiako (1990), cited by Muroi and Sasaki (1997: 186) explains:

> this company has been making use of sex tours as a reward to sales workers. Although these tours do not attract workers as before, since they have got used to them and their wives are against these visits, the company cannot find better

rewards than giving an opportunity to go on sex tours. These tours are both physically and psychologically satisfying.

In a nutshell, tourism has the power to unexpectedly bring people from different cultures together, which occasionally leads to serendipitous encounters between the visitors and the visited. Such encounters can sometimes result in enduring relationships. This visitor–visited model takes into account the varying roles that the stakeholders of the tourism industry play in CSW–tourist sexual encounters (Figure 1.2). These encounters are temporary relationships with finite time limits, making it emotionally safe for many (potential) clients/tourists, as there is little chance of rejection. They are liminal relationships occurring outside one's social norms which allow the concerned individuals to explore different types of relationships sometimes with different people (Figure 9.2). It is an opportunity for a complete break from one's normal behaviour, which can be both liberating and reinvigorating. Thus, tourism provides an environment free of the constraints and inhibitions of home, providing increased opportunities for engaging in sexual relationships (see 'Setting the Scene'). The tourist seems to have escaped his or her social roles. In other words, a traveller gets an opportunity to express things he or she would otherwise suppress. Consequently, a teacher, for instance, can temporarily become someone else while in a destination away from home.

Commercial Sexual Service Consumers in Kenya

Commercial sexual service consumers in Kenya form a highly heterogeneous group with a large range of incomes and diversity in lifestyles. They may be women or men, homosexual or heterosexual, constituting the middle class or the upper class of their society. For the purpose of a structured discussion, this section groups these CSWs' clients into male and female sex tourists.

Male Sex Tourists

From a general point of view, gender roles have been affected by transformations of human practices and relations through modern consumer culture. Up to the mid-twentieth century, man was considered the producer, in the workplace – public domain – while the woman was the consumer, in the home – private domain (McLeod 1982; Pruitt and LaFont 1995; Ndune 1996; Cabezas 1999; Dahles 1999). In the Western world, a new model of gender relations is replacing this traditional male provider–female consumer paradigm – equal sharing of responsibilities, and rights to providership and sex. In the process, women have slowly increased their share of economic power, thereby making them less economically dependent upon men. As a consequence, women's share of relationship power increases.

Some male sex tourists are resentful of this perceived power of women (Mathews 1978; Burton 1995; Pruitt and LaFont 1995; Cabezas 1999). They fear

Western women's ability to reject their sexual advances. Jan (fictitious name), a regular tourist to Watamu (Figure 6.1) from the Hague, argued that Dutch women demand too much from men:

> It is unbelievable ... back at home, the girls I fancy do not even look at me and the ones that fancy me tend to be sort of fatter and older. I do not like them. However, in Kenya, beautiful girls love me. They are all over me. I feel like a rock star while in Kenya. My girlfriend is so beautiful, it puts me to shame really .... Women in Holland are bitchy – they are too demanding. I am only a 43-year-old painter, and I am doing very well in life. But Dutch girls, they want someone with a prestigious and well-paying job. They do not want someone like me. They want a lawyer or a doctor or something of the sort. They want to move up in society, and I cannot blame them .... On the contrary, Kenyan girls do not expect so much. If you take a Kenyan girl out for dinner, she's grateful, whereas a Dutch girl, she's grateful but she wants more ....

Consequently, one is tempted to conclude that some Western male tourists who resent women's perceived power at home turn to women in Third World tourism destinations. This gives them a chance to escape the pressures from their wives, girlfriends and/or female colleagues as they continue with their relentless pursuit of gender equality of all sorts (see, for example, Pruitt and LaFont 1995). In contrast, female CSWs in Kenya neither challenge nor demand anything much from male sex tourists. Jan concluded:

> You just need to make sure that you are well prepared for evening sexual encounters, do some shopping for her, give her some pocket money ... and you are set. For your information and for your future reference, I am not leaving this country [Kenya]. I have decided to settle here. I feel really appreciated here ....

Bill (not his real name) another sex tourist in Shimoni (Figure 6.1) and a high school teacher from Wisconsin, said: 'I like coming to Kenya because women here treat me like a man, a desirable personality.' He added: 'In the US there are ten men for every girl while in Kenya there are ten girls for any man who appreciates good (sex) life, and all of them eager to please." As if to confirm the foregoing, Oliver (pseudonym) a sex tourist from Yorkshire, bragged about how he had 'sex with the two young girls'[1] he was playing beach volleyball with, then the girls washed his feet on the beach, put sun tan lotion on his back and cleaned his rented room, all for US$50.

---

1 In Kenya's coastal region, a commercial sexual encounter involving one man and two women is referred to as 'one-by-two', while 'two-by-one' refers to sexual services involving two men and one woman. A sexual encounter involving two men and two women is called 'double-double'.

Likewise, based on interviews with Thai female CSWs, Cohen (1982) shows how economic gains made by the women in these kinds of interactions, both short- and long-term, are offset by significant social and psychological costs. This type of CSW–tourist relationship is a mirror image of the dependency of African/developing countries on the developed nations (see Chapter 3; Figure 9.2).

Female Sex Tourists

Western women are also turning to Kenyan men to challenge some of their own men's traditional roles or to live out their fantasies. Such female sex tourists affirm their sense of womanliness by being sexually desired by other men (Figure 9.2). In their home countries, these women may be stigmatised for having either illegitimate or casual relations with multiple sexual partners. A female tourist in Malindi observed: 'Here one is allowed anonymity to enjoy liaisons with men as one desires, all out of view of nosey neighbours and friends. It is freedom in its strict sense.'

Consequently, Kenya's coastal region, like Negril in Jamaica, Kuta in Bali, Boca Sosua in the Dominican Republic and some resorts in the Gambia, is becoming a popular destination where Western middle-aged women come in search of what they call the 'African banana'.[2] This is definitely a contradiction of classical analyses of sex tourism which do not allow for the possibility of women travellers as buyers of sex. This is because the users of commercial sexual services are typically male (Mathews 1978; Pruitt and LaFont 1995; Jallow 2004). Many researchers and theorists share this assumption (see, for example, Fanon 1966; Delacoste and Alexander 1988; Smart 1992; Mwakisha 1995; Günther 1998; Cabezas 1999; Kempadoo 1999; Sindiga 1999; Trovato 2004).

Female romance tourism To understand the manifestation of the phenomenon of female romance tourism in Kenya, a survey was conducted in Malindi (Figure 6.1) during March–May 2006. Data was gathered from a questionnaire, in English, administered to an opportunistic sample of 68 female tourists. In addition, the author conducted in-depth structured interviews with the respondents. The questionnaire sought two distinct kinds of information. Part I covered issues of demographic characteristics (age, marital status and origin), while Part II dwelt on consumption of commercial sexual services by the respondents while on holiday in Kenya. Tables 5.1 and 5.2 detail the origins and age structure of the female sex tourists in Malindi respectively.

Overall, 22 per cent of the respondents were Germans, while 19 per cent were Italians (Table 5.1). The majority (61 per cent) of the female sex tourists visiting Malindi were 46–50 years of age (Table 5.2). Seventy per cent were re-visiting the destination. Chi-square analyses were carried out to establish the relationship

---

2   'African banana' is (female) sex tourist jargon for the African male's sex organ. It is also referred to as 'African bamboo'.

Table 5.1 Origin of female sex tourists in Malindi ($n = 68$)

Origin	%
Germany	22
Italy	19
Netherlands	15
Norway	13
UK	12
US	11
Others[a]	8

Note: [a] France, Ireland and South Africa.

Table 5.2 Age groups of female sex tourists in Malindi ($n = 68$)

Age group	%
31–35	3
36–40	8
41–45	19
46–50	61
51–55	6
56–60	3

between selected demographic characteristics of the female sex tourists ($n = 68$) and consumption of commercial sexual services by the respondents while on holiday. Tests for association between age, origin and marital status revealed no relationships in all the three cases with $\chi^2_{(3)} = 5.64$ at $P<0.100$ (age by consumption of commercial sexual services), $\chi^2_{(4)} = 4.87$ at $P<0.100$ (origin by consumption of commercial sexual services) and $\chi^2_{(4)} = 4.20$ at $P<0.100$ (marital status by consumption of commercial sexual services). These values indicate that consumption of commercial sexual services by the female sex tourists while on holiday (in Malindi) was not dependant on their age, origin or marital status. This observation can be attributed to the fact that none of the respondents was on an organised sex tour. Moreover, 94 per cent of these female sex tourists were independent travellers.

Further, the survey found that 66 per cent had the intention to engage in sex with local men at their destination before departure from their home country. Interestingly, 84 per cent of the female tourists who had a prior intention to enter into a sexual relationship with the local men was almost equally distributed among the top six countries generating female sex tourists to Malindi – Germany (19 per cent), Italy (18 per cent), the Netherlands (17 per cent), Norway (16 per cent), the UK (15 per cent) and the US (15 per cent). The rest of the female sex tourists came from France, Ireland and South Africa (see Table 5.1). Ninety-one per cent of them knew about Kenya being a 'safe' destination for women on 'romantic tours' (a term used by the interviewees to refer to their visit to Malindi), either through their past experience in the country (54 per cent) or through recommendations by their friends or colleagues – word of mouth (46 per cent).[3] This is another proof that Kenya is not being marketed as a sex tourism destination by international tour operators or travel agents. Nevertheless, by the time of this study (March–May 2006), 59 per cent of the total respondents had already entered into one or more sexual relationships with local men while on holiday in Kenya.

Although 60 per cent of the respondents admitted to certain 'economic elements' to their liaisons, they did not perceive their encounters as CSW–client transactions, nor did they view their sexual partners as CSWs. The 37 per cent of the total female tourists who admitted to coming to Kenya for sex believed that they were helping the local men, and to some extent the local economy, by giving them money and gifts (see, for example, comments by Jean-Baptiste quoted in Chapter 1). Interestingly, all of the female tourists who participated in this mini-survey felt that they were not sex tourists. This confirms our earlier observation that very few tourists would define themselves as sex tourists (see Chapters 1 and 2).

Penetration of male domains through romance tourism   Women have traditionally used travelling as a way of masculinising their identities, rather than as a way of affirming their femininity (Delacoste and Alexander 1988; Pruitt and LaFont 1995). Today, some female sex tourists are travelling in order to penetrate traditional male domains, claiming traditional male powers to reaffirm their femininity (Maurer 1991). It is important for many female sex tourists to affirm their sense of 'womanliness' by being sexually desired by men. When asked why she visits Kenya, Pat (fictitious name) a 42-year-old Irish woman, responded:

> I love the Kenyan climate and the indigenous people .... I especially love Kenyan men. They are skilfully flattering. They make you feel like a woman. They are loving and artfully work your emotions. I pay for their drinks and meals for their 'tender night care' [*sic*]. Yeah, I want fun and good sex ... which is lacking back at home. Long live Kenya! Long live Kenyans!

Women who feel rejected by men in the developed countries for being 'sort of fatter and older' find that in Kenya all this is suddenly reversed (Kibicho 2004b). There, they are 'romanced', appreciated and 'loved' by men. They are regarded as being more voluptuous and sexy (Pruitt and LaFont 1995). In consequence, the female sex tourist has become besotted with the local man, for who else will

---

3   Word-of-mouth communications play an important role in forming tourists' expectations. Tourists tend to rely more on personal sources of information than on non-personal ones when choosing among (sex) tourism destination alternatives. They view word-of-mouth information as unbiased intelligence from someone who has been through the (sex) service experience.

put his hands around the rolls of fat that surround her waist and tell her, looking deep into her eyes, that she is sexy and that he loves her (Phillip and Dann 1998)? Kenya's sex tourism thus allows some Western women to sexualise their bodies in ways that would be difficult to achieve back home, and to be desired by highly desirable men. In sum, sex tourism thus offers women one of the few legitimate avenues to self-actualisation (see Figure 9.2).

The socio-economic status enjoyed by these women provides them with independence and security that translates into power and control of their relationships with the local men (Mathews 1978; Pruitt and LaFont 1995). Moreover, half of the female tourists interviewed during this study reported that they enjoyed the power that money gives them over men while on holidays in the visited destinations. Consequently, one may correctly argue that with the increased socio-economic status of women globally, there is an undeniable likelihood of there being more female sex tourists, and thus more male CSWs serving females in the future (see also Chapter 8).

Although this was a non-random sample, and thus scientifically irrefutable generalisations cannot be advanced on its basis, what one can conclude is that some women travel to Kenya for sexual services, just as some men do. Moreover, it seems that female sex tourists are very similar to male sex tourists in terms of their attitudes and motivations and the narratives they use to justify their behaviour (see Burton 1995; Pruitt and LaFont 1995; Kempadoo 1999; Kibicho 2004b).

Before going on to consider the role of information technology in the overall development of sex tourism, it will be useful to discuss the general motivations of sex tourists in Kenya.

Motivations of Sex Tourists in Kenya

It is difficult to analyse the tourism-related sex trade adequately without examining socio-cultural features of commercial sexual services consumers. Why do they buy these services? Are there specific reasons? Certainly, there are many possible reasons for seeking sex in another country or region. Among them are the more relaxed laws governing commercial sexual services, cheaper sexual services, greater anonymity, and sexual adventurism. The remainder of this section concentrates on a detailed examination of these reasons as identified by some sex tourists in Kenya's coastal region.

Relaxed Laws on Commercial Sexual Services

Occasionally, relatively relaxed laws regarding commercial sexual services are accompanied by less rigorous enforcement of laws governing sexual behaviour or sex trade in general. Consequently, potential sex tourists develop a feeling of being uninhibited in engaging in many sex-related activities that are not tolerated in their home countries (see also Chapter 2). Countries with relaxed laws on sex trade thus provide conducive environments for the growth of sex tourism.

In many tourism destinations in the developing countries, for example, sex tourists indulge themselves in sexual fantasies of all kinds without fear of social stigmatisation or harassment by law enforcers. The contradictory laws on sex trade (see Chapter 2), complemented by their poor enforcement, in Kenya serve as a key motivator for sex tourists to visit the country. This also includes a significant percentage of child sex tourists, who visit almost all tourism destination areas in the country.

Cheaper Sexual Services in the Developing Countries

In many cases, comparatively cheaper sexual services motivate (sex) tourists to visit a given destination (see Rogers 1989; Afrol 2003; Jallow 2004). This cheapness is typically due to tourists travelling from an economically wealthy country to a poorer one. Thus, from the tourist's viewpoint, sexual services in the developing countries are less expensive (see Figure 9.2). For instance, Pat, the Irish female tourist cited earlier in this chapter, observed: 'in Kenya you have a beautiful tan, you feel gorgeous, you are treated like royalty ... everything, including unlimited sex, is easily available and cheap' (see also the comments by a German male sex tourist cited in Chapter 4). Consequently, the relationship between sex trade and tourism is particularly strong in countries where 'low-cost exoticism' in both tourist services and commercial sexual services is one of the main attractions (Cazes 1992; Ackermann and Filter 1994; Burton 1995; Hall 1996; Cabezas 1999; Kempadoo 1999).

Surprisingly, in addition to low costs, some tourists find having sex in the tropical warm conditions more enjoyable. For instance, a 44-year-old Italian female sex tourist observed:

> I like having sex on the beach under the ... tropical sun. It is just wonderful to have these sexually untiring local men on you under the burning sun. You know, Malindi's hot sun .... But of course, what is more enticing to me is the sex, not the sun. ... It is just a complementary.[4]

Anonymity While Away from Home

Regardless of the 'pull factor' of unlimited concentrations of 'sex facilities', there are associated 'push factors' guiding international sex tourism. There is the push of the exclusion from 'normal' society and the consequent need for the reassurance of the open and secure company of other sex tourists. Sex tourists are, in large part, able to be themselves only in the 'tourist space'. This is a space/environment that is limited and 'artificial', but is often the only one where sex tourists can

---

4 See also Pat's comments above.

be themselves. Thus, sex tourists use time–space strategies to separate the performance of their 'sex tourist' identity from the performance of their identity as fathers/husbands, mothers/wives and girlfriends/boyfriends.

In addition, many sex tourists will choose to travel in search of an anonymous and thus 'safe' environment in which to be sex tourists. They choose to travel for sexual encounters they wish to conceal from others. In other words, they feel less inhibited outside the constraints of their home countries, and may be attracted by what they feel to be less restrictive social taboos in other countries. They are thus free to engage themselves in sexual activities otherwise socially 'unacceptable' back home. Such sexual activities include having multiple sex partners, having sexual encounters with people from certain backgrounds (race, social status) and having sexual liaisons with minors. In brief, the holiday gives a considerable opportunity to escape into a world of fantasy and indulge in kinds of behaviour generally frowned on at home.

Consequently, some tourists feel that they can discard their moral values while away from home and do not feel accountable for their behaviour and its consequences (ECPAT 2002; Afrol 2003; Jallow 2003). In other words, tourism represents an escape from one's normal life, a chance to become another person temporarily, or to engage in another lifestyle (Dewailly 1999). Thus, a tourist will behave differently while on holiday. This temporary dislocation of the tourist from normal life explains why tourism, love, romance and sex are closely linked (see, for example, Bauer and McKercher 2003).

Stereotypical Beliefs about 'Others'

Some tourists believe that people from certain origins are more sexually active (Figure 9.2) (Fanon 1966; Brown 2000; Hall 1996; Kempadoo 1999). For example, Jan, the Dutch tourist quoted earlier in this chapter, reported that the difference between local Kenyan girls and those of his race created 'a strong psychological stimulation'. A 38-year-old German female tourist from Bonn added:

> this is my first time to take a holiday explicitly with a sexual encounter as my principal goal. I was keen to find a Maasai man as I had read and heard that they are endowed sexually [*sic*] ... and today I can confirm it – they are hot. They are sex *morans* [champions]. Their women are lucky ... certainly, they are not as sexually starved as we are back at home ....[5]

Of course, it can be argued that the spatial and cultural detachment of the tourist from the CSW reduces the guilty conscience on the part of the commercial sexual service consumer. Moreover, sexual imagery used in the marketing of some post-colonial tourism destinations tends to be a continuation of Western representations

---

5   The Maasai are one of the 42 communities in Kenya. These people are also found in northern Tanzania.

of sexually available and subservient females dating from the nineteenth century (Hall 1996). With regard to Africa and Africans, McLintock (cited by Hall and Tucker 2004: 11) observes that:

> Africa was established as the quintessential zone of sexual aberration and anomaly in European lore as the very picture of perverse negation that declared Africans to be proud, lazy, treacherous, thievish, sexually hot and addicted to all kinds of lusts ....

However, according to Hall and Tucker (2004), McLintock is of the opinion that any focus on race or gender as defining categories for a sense of self are insufficient. Nevertheless, it would not be overstating the case to say that any tourist sharing such stereotypical sentiments is highly likely to reason and behave like the German tourist quoted above.

On a different note, negative stereotypical attitudes, based on inaccurate or insufficient information, result in hatred. This is more so when one's fantasies are not satisfied. This eventually 'destroys one's sense of values and his/her objectivity causing him/her to describe the beautiful as ugly and ugly as beautiful, and to confuse the true with the false and the false with the true' (Martin Luther King Jr, 1929–1968).

Sexual Adventurism

The desire on the part of tourists to try something new – sexual adventurism – is another motivating factor in sex tourists seeking commercial sexual services in a foreign tourism destination (Fanon 1966; Mathews 1978; Günther 1998; Kempadoo 1999). This confirms the basic need of tourists for variety, novelty, and indeed fantasy. Some sex tourists who have visited other sex destinations (for example, Cambodia, Thailand and Vietnam) feel the need to sample commercial sexual services from emerging destinations like Kenya. To qualify the foregoing, consider the following comments by a female sex tourist from Lille, France, in her early forties:

> I have been to the Philippines and then to Thailand. Although I do not regret the quality of the sexual services I received there, I decided to taste a new product, the 'African banana' .... I have never experienced such a sexual experience, and you can quote me on this .... Good God, these guys are good. I am definitely coming back next month. I wish I'd discovered this sex magic when I was young. You know, I was a sex bomb ....

Impulse Sexual Service Consumption

As noted elsewhere in the text, some tourists do not plan in advance to purchase sexual services while in Kenya (see also Cohen 1982, 1988a). In many instances, it is the CSWs who approach the potential clients, who are enticed by the low prices for sexual services (see Table 9.3). One example is a 44-year-old Norwegian female tourist who said:

> To be honest with you, I had not planned to have a sex relationship with a local man. It all started at the beach. This guy [referring to her sex partner] sold me a souvenir then proposed a very cheap safari to Tsavo West National Park. We spent the following day in the park together. Since then, we have become [sex] partners .... Definitely, this cannot be called sex tourism. We are simply helping one another. He helps me to see and discover Kenya, while I help him with some pocket money as a sign of appreciation. Having sex with him is just a side thing ... to keep ourselves busy at night. Of course, I enjoy it. I am a normal woman ... with normal sexual feelings. But, I am not a sex tourist. I am against sex tourism. It degrades the local people. It is disrespectful to local cultures.

Technological advancement in the field of tourism marketing has also played a significant role in the expansion of sex tourism in Kenya, in Africa and in the world in general. Towards this end, the Internet revolution and virtual reality ingenuity are not blameless. This is the focus of the next section.

Information Technology and Development of Sex Tourism

The Internet as an Effective Tool in Promoting Sex Tourism

The Internet has become the most effective and fastest way to transmit information around the world (Gould 1995; Oppermann et al. 1998). It is a highly effective tool for promoting the development of the tourism industry in general, and sex tourism in particular (Kohm and Selwood 1998). Those seeking sex tourism destinations on the Internet have a bewildering array of options, as sex is a major item for sale on the Web.

Logging in provides a potential sex tourist or cyberspace sex tourist with immediate information on where to go, how to get there, when to go there, and what one needs. In other words, Internet portals are today embracing lifestyle segmentation to assist their clients in creating the appropriate tourist product to complement their lifestyles. There is basically something to satisfy the tastes of all sex tourists in cyberspace, 'whether it is simply voyeuristic viewing of "naughty" images or satisfaction of the most aberrant (sexual) needs' (Kohm and Selwood 1998: 126). Some of sexual services available in cyberspace include 'hot chat', phone sex, bondage and discipline, X-rated videos/DVDs and erotic pictures. Sexual aids are also available on the Web.

Further, the Internet has opened up a new virtual world of sex tourism (Oppermann 1998; Oppermann et al. 1998).[6] Websites on the Internet provide an international forum where individuals can promote and sell sex tours online, sometimes advertising packages for travellers complete with airfares, hotels and directions to local brothels. As a consequence, sex and travel on the Web are coming closer to mimicking conventional sex tourism experiences. One can thus have sexual titillation and satisfaction in almost any imaginable form through surfing the Web from the comfort, convenience and privacy of the home or the office. Of course, not 'all of the senses can be stimulated, but enough of them can be sufficiently activated allowing the imagination to complete the experience' (Kohm and Selwood 1998: 123). Moreover, the Web allows its users to disguise their identity and to indulge in fantasies far more easily than in real situations and relationships.

Some virtual sex tourism destinations, such as <www.dreamsex.com>, <www.pussy.bahnhof.se>, <www.bekoame.or.jp>, <www.sexmall.com> and <www.intersex.com>, offer a full range of sexual services. They also feature sexually oriented nude pictures. SingleTravel.com (<www.singletravel.com>) provides information on brothels as well as various sex destinations around the world. It is of interest to note that new sex sites are proliferating and existing ones are constantly expanding their products and services as new technologies and clients become available (Kohm and Selwood 1998; Dewailly 1999; Kempadoo 1999). From a general viewpoint, these sites are essentially travel guides for conventional sex tourists.

Virtual Reality: An Alternative for Sex Tourists?

The success of tourism developments depends largely upon the image which society creates of them (see Chapter 9) (Gould 1995). In other words, a geographical space gains touristic importance if its resources coincide with society's tastes and expectations. Words are used to create these images. These are supplemented by virtuality, which has completely revolutionised the art of formation and development of (sex) tourism destination images conjured in the minds of (sex) tourists. Virtual Reality consists of three elements: (1) visualisation of the tourism destination through simulation, (2) immersion in the destination and (3) interactivity with elements of the destination. It offers potential tourists still and mobile images which can be either real or unreal. Real images are accessible, while unreal ones are inaccessible, as they are a reconstruction of reality. Virtual reality blurs the difference between reality and imagination. In any case, according to psychologists, transitioning to actual behaviour kills fantasy (Kohm and Selwood 1998).

---

6 The most commonly used virtual imagery tools today are CD-ROMs and video cassette tapes.

Virtual reality allows interaction and reaction through avatars and TFUI (touch-and-feel user interface) and three-dimensional experiences courtesy of VRML (Virtual Reality Modelling Language). These include interplanetary trips or voyages to fantasy worlds which are inconceivable in reality – cyber-tourism or surrogate tourism.[7] Simply strapping on a body suit and plugging into a virtual reality program can transport a (cyber)sex tourist to the sights, sounds and sensations of, say, a RLD in Amsterdam. It has opened vast new fields of application, which have hitherto been perceived with a degree of uncertainty (Günther 1998; Oppermann et al. 1998; Kohm and Selwood 1998). The participant experiences sensations as if he or she is at the centre of the action. The emotional reaction to these virtual experiences is physically real, as expressed by shouts and instinctive body movements. Although from an academic standpoint one might argue that these (sex) tourists are not having an authentic experience, it seems of little consequence to those involved (see also Chapter 4).

It is undeniable fact that the perception conveyed in the electronic media concerning a specific tourism destination can be a decisive element in its visitation levels. The visual appeal of a tourism site tends to be enhanced by motion pictures or television/documentary series (Butler 1980; Launer 1993; Ackermann and Filter 1994; Dewailly 1999). Likewise, the creation of virtual sex sites will increase sex tourist flows in real sex tourism destinations. This is more so as virtual reality allows manipulation of a destination's attributes, thereby enhancing potential tourists' interest in it (see Gould 1995). It is constantly perfecting the sensory interaction that it brings into play. Under these circumstances, therefore, taking virtual sex vacations might be in the offing. In any case, through virtual sex tourism, a (cyber)sex tourist will avoid loss of time, long queues, potential violence, risks of accidents, language problems, the threat of STDs (associated with conventional sex tourism) as well as other cumulative expenses.

However, virtuality does not replace direct social and cultural experience, or memorable interaction with the physical environment. Consequently, one is tempted to conclude that virtual reality will not replace real sex tourism experience to the point by becoming its perfect substitute (see, for example, Dewailly 1999). Nevertheless, a combination of the real and the virtual can create a growing diversity and complexity which ultimately adds to a destination's richness in terms of its tourist product and the range of alternatives for its clientele. Moreover, the very nature of sex tourism as an experience lends itself perfectly to virtual reality.

Like the conventional sex trade, the expansion of cybersex[8] is facing obstruction from the usual quarters – government regulations (Kohm and Selwood 1998). In fact, some cybersex tourists have been indicted for distributing and receiving pornographic materials, especially those involving children, in a number of jurisdictions. However, effective policing of transactions on the Web

---

7 Surrogate tourism is the quintessential post-modern tourism experience – everything is connected, yet nothing truly fits together.

8 Cybersex mainly entails anonymous Internet romance.

has been rendered difficult by the high volume of traffic. On the same strength, website owners are protecting themselves by requiring users to acknowledge their willingness to view 'sexually offensive' material before gaining access.

In the mean time, the debate as to the real impact of virtual reality on (sex) tourism continues. On one hand, opponents say that virtual reality may depress demand for real (sex) tourism. This argument is based on the fact that virtual reality is increasingly rendering homes and offices central and secure bases for sexual and other leisure activities. On the other hand (as noted above), proponents are of the opinion that virtual reality will simply whet the appetite for more travel through enhanced exposure to, and awareness of, the (sex) tourist product, as virtual reality only serves as an advanced medium for (sex) tourism promotion. All in all, there is no doubt that the Internet and virtual reality have created a paradigm shift across the (sex) tourism industry, with an impact both upon the way the industry is organised and operated, and upon how (sex) tourists behave.

All these motivating factors aside, however, socio-economic considerations play a crucial role in the expansion of the tourism-oriented sex trade in Kenya and other destinations in Africa and in the developing world at large. This forms the core of the discussion in the following section.

Economic Enablers for Sex Tourism in Kenya

Sex tourism, a sub-sector of Kenya's prosperous tourist economy, is a significant industry and a major employer for many CSWs (Kibicho 2006c, 2007). This is not forgetting those in other sectors of the economy indirectly related to the sex industry, such as taxi operators, tour operators, hoteliers, beer makers and healthcare providers. The role of this sub-sector is obvious considering the number of CSWs visible in the streets, bars and hotels openly soliciting customers. Although the number of sex tourists and CSWs in Kenya is unknown due to the illegal status of sex trade (see Chapter 2), what is known is that a quarter of the country's investments have been made in the tourism industry (Kenya 2003, 2004; Kibicho 2005c). This makes the tourism industry one of the most flamboyant and dynamic economic sectors in Kenya (see Chapter 3).

Demand for commercial sexual services in Kenya, and indeed worldwide, can be examined from both economic and social viewpoints. From an economic standpoint, for example, CSWs earn considerable profits from sexual services. Sex workers also charge foreign tourists more than domestic ones for similar sexual services. In addition, pimps earn huge profits for linking CSWs with clients (see Chapters 6 and 8). From a social viewpoint, women and girls are never given equal opportunities to their male counterparts in many societies (McLeod 1982; Middleton 1993). Such disparities range from access to education to job opportunities. Thus, inequalities in the domestic domain intersect with inequalities in purportedly gender-neutral institutions of markets, state and community to make gender inequality a society-wide phenomenon. A combination of these factors in the Kenyan patriarchal societies, for instance, leads many women to the sex trade. Poverty – and gender inequality – therefore have to be tackled at the societal level as well as through explicit interventions tailored to addressing specific forms of disadvantage. In simple terms, promoting gender equality in the wider African society – or reinforcing inequality – can have effects that cut across class and social status. The next section examines how poverty contributes to the development of child sex tourism in Kenya.

Nexus between Poverty and Child Sex Tourism in Kenya

Child sex trade is a result of a combination of complex socio-economic factors, among which poverty is a main one (Chapter 9). Kenya's children are introduced into the sex industry in various ways. Some are lured away from broken homes by recruiters who promise them jobs in the city (see Chapters 1 and 6). Once away from their families, these children are forced to have sexual liaisons with multiple clients. Others are introduced into sex trade by their parents in a desperate attempt to earn some extra money (De Kadt 1979; Ndune 1996). Economic austerity, rising unemployment, high inflation rates and growing poverty create intense desperation, forcing parents to sell their children in exchange for basic necessities such as food and shelter. In such circumstances, the fact that the child will be sexually exploited is either ignored or misunderstood (Oppermann et al. 1998; Kempadoo 1999). One example is a family in Kilifi (Figure 6.1) which exchanged its 6-year-old daughter for a set of cell phones worth US$99 with a Norwegian couple. This deal was disguised as adoption. With the help of corrupt bureaucrats at the Immigration and Adoption Office in Mombasa, the couple officially adopted the girl through a successful application to the Attorney General's Chambers in 2000 (Kibicho 2004b). Four years later, the girl's father launched an appeal at the local law courts after realising that the girl was being used in pornographic film production in Oslo (source: field notes). The case was later determined in favour of the Norwegian couple.

Stories of foreign tourists buying expensive gifts, houses and other properties for their 'children friends' are rampant in Kenya. Surprisingly, some parents remain silent when their children bring home money or other forms of gifts from their tourist friends. Further, these parents are ignorant of their children's activities and where they spend their free time. With the aim of continuing in this dream world of a relatively easy life well marked by easy money and 'good times' – juxtaposed with the harsh reality of economic rationalism, high unemployment rates and abject poverty – such children constantly frequent the neighbouring tourist enclaves. After some time, they abandon their family/social obligations as they try to copy the tourists' social and moral norms, which leads to disorientation from their own values (Figure 9.2). This in turn results in deterioration of relationships within family networks, as well as within social groups. Under such conditions, these innocent and unsuspecting children become vulnerable targets for child sex tourists. It should be remembered that Kenyan children are taught not to suspect others, especially adults, unless they instinctively feel threatened (Ndune 1996; Mungai 1998; Kibicho 2005a). In any case, parents and members of the larger community share responsibilities for teaching youngsters societal values as well as individual obligations. In brief, a Kenyan child belongs, or probably belonged, to the whole community.

In other instances, abandoned children or those from families with a history of drug abuse or alcoholism or from poor backgrounds are found wandering in the tourism destination areas, especially on the beaches. Sometimes they approach tourists in their daily begging expeditions. Some of the tourists approached take this opportunity to develop a child–tourist friendship. To illustrate this process, consider Joan's (pseudonym) story:

> I started 'working' at Jommo Kenyatta Public Beach (Mombasa) at the age of eight years. I begged for money from both Kenyans and foreigners. Five years later, I met Steve [not his real name], an American tourist from Florida. During the first days, he would invite me for lunch in his hotel and then give me about five US dollars to take home. One day he invited me for dinner. We had a super dinner accompanied with a lot of wine. He then requested me to accompany him to the beach. It was about 20H00 [local time] and thus darkness had already set in. While on the beach, he started caressing me … you know, touching me everywhere [she smiles]. At first, I got extremely scared …. It was my first time to be touched like that by a man. He assured me that he loves me so much that he could not do anything harmful to me.

> The next moment, I felt him inside me, although with a lot of pain [she laughs loudly]. That was the beginning of the whole thing – the commencement of my sex encounters with men for money. We could do it [have sex] on daily basis … in his hotel room, in the discothèque clubs, in the car parks, in his car, on the beach …. Of course, he made sure that I was well fed. In other words, he had solved one of my major daily problems – hunger.

> He left Mombasa after five months. Thus, I had to look for another *mzungu* [Kiswahili word for a white man]. Yes, I tried African men on several occasions. Nevertheless, I can assure you that they are very mean … in fact stingy. Occasionally, some of them had sex with me, but then refused to pay for my services. Therefore, I decided to specialise in *wazungu* [plural of *mzungu*], and as you can see, I am doing quite well. Probably better than you [referring to the author]. So, to cut a long story short, Steve introduced me to this game [sex trade] when I was 13 years old …. Today, it is the only thing I know how to do best and thus I have no immediate intention to quit the profession. Or, to put it in another way, I am not jumping off the diving board [leaving the sex trade] until I am pretty sure that there is some water in the pool [better economic alternatives].

Another example is Nick (fictitious name), who was 12 years old at the time of this interview, June 2005. He has been used in pornographic films. He was recruited by an older boy, a beach boy, for an Australian pornographic video. The Australian tourist was a regular visitor to Watamu, a local tourist resort 25 kilometres south of Malindi town (Figure 6.1). He was staying in a rented house. The older boy tricked Nick and another younger boy (10 years of age at the time of data collection for this study) and took them to the tourist's house. He promised the boys that if they co-operated, he would request his *mzungu* friend to organise for them a visit to Australia. They were both excited about the offer. It is worth mentioning that in many instances, the older boys socialise the younger ones into oral sex before they are introduced to the potential child sex tourist. This socialisation process involves performing oral and sometimes anal sex acts. Nick narrated the rest of the story:

> The room was lined with gigantic mirrors. The *mzungu* forced us to touch each other's private parts, 'suck' one another and then have sexual intercourse in front of huge cameras. All along, they [the tourist and the beach boy] would first demonstrate to us what we were supposed to do. First, our friend [the beach boy] 'sucked' the *mzungu*, and then they had sex. The tourist told us that if we want to become white [*sic*] and thus have money like him, we should start 'sucking' him regularly like our friend [the beach boy]. After five hours, each of us was given 20 US dollars plus a bag of gifts – sweets and chocolate. In my case, my father took all the money. He did not even take me to the hospital, for I was having a lot of pain, especially when walking, as the *mzungu* had sodomised me ....

Mercy (pseudonym) tells a 'success' story. Coming from Central Kenya, she got into the sex business at the age of 14. Three years later, she met a Canadian business magnate with whom she spent four weeks travelling the country. According to her, by the end of this period she had 'succeeded' in her task – told her companion the story of her sick mother and two younger brothers who needed money to go to school. She noted:

> Today, at 18 years of age, I am no longer poor. I invested the money I got from my friend wisely .... I export curios to Europe, especially Italy and Spain. I built a good permanent house for my mother, and I pay for my brothers' education. To crown it all, I own a five-bedroomed bungalow with a spacious garden in the outskirts of Mombasa City. By your standards, is this not being successful?

However, sexual exploitation of children in Kenya is not the preserve of sex tourists or the tourism industry. As noted in Chapter 1, many Kenyans, including relatives of the children concerned, actively participate in the exploitation process. To demonstrate this, let us consider 16-year-old Doris (not her real name), who had to abandon education at standard/level 5 to seek 'jobs' in order to help her poor parents support her siblings. Even then, the situation at home was still pathetic, marked by irregular meals. Consequently, when her uncle suggested that she accompany him to Malindi (Figure 6.1) to work at his curio shop, the family readily accepted the idea. A few days later, the two arrived in Malindi, where she worked as a curio shop attendant as well as doing some unspecified errands for her uncle for two months. Doris was shocked when one night, her uncle went to her room, as she was living in one of the bedrooms in her uncle's house, and demanded to have sex with her. He argued that sex was part of the deal they had negotiated with her father. This sexual exploitation continued for several months, until one of Doris's friends offered her an alternative place to stay. However, this relief was only short-term, as she was forced to quit her job at the curio shop because the uncle was incensed by her decision to move to her friend's apartment. At the same time, and with the same vigour, Doris's friend started demanding a contribution to the house rent and daily spending, especially food costs. Doris reported:

> Although I was jobless, my friend threatened to throw me out of her apartment. She encouraged me to join her business [sex trade] if I wanted to earn easy money and live a good life. She argued that it is better to earn something from sexual services than to give it out for free to uncaring men like my uncle. From, then on we became business partners, as we shared all the domestic expenses. One year later, we moved to a bigger apartment within the town centre. This is where we operate our business from. We entertain clients from all walks of life ... ambassadors to taxi drivers; tourists to non-tourists ....

The above example shows clearly how a trusted caregiver can exploit the vulnerability of an innocent girl, thereby betraying her trust. But apart from the tourism-oriented curio shop Doris was working in, her predicament has nothing to do with the tourism industry. Or does it? The question is: How many Dorises do we have out there? However, this question is beyond the scope of the current section. The critical finding here is the existence of non-tourism-induced child exploitation in Kenya. Thus, the tourism industry should not be blamed for all sex-related ills in a tourism destination area. Consequently, government policies aimed at combating such social challenges should address *all* socio-cultural, economic and political spheres of societal needs.

Last but not the least, it is necessary to underline the fact that many of the child CSWs are beaten, tortured, or in extreme cases, murdered (Caplan 1984; Brown 2000; ECPAT 2002; Afrol 2003). Some of them contract STDs, with others dying of HIV/AIDS-related illnesses. To endure these situations, albeit momentarily, many of them turn to alcohol and/or drugs. Those who are lucky enough to eventually leave the sex industry often have emotional instabilities, a sense of being worthless and fundamentally bad, and are incapable of loving or trusting or forming healthy intimate relationships. All said and done, however, when dealing with child sex tourism, the focus should be on addressing the deep-rooted structural inequities within the general development paradigm which exacerbate poverty and constrain attempts to alleviate poverty.

Beach Boys: The Masked Front of Kenya's Sex Tourism?

As mentioned in Chapter 4, sex trade in Kenya differs from prostitution in Western countries (see also Sindiga 1995; 1999; Ndune 1996; Kibicho 2004a, 2004b, 2005b). For instance, there is neither a formal network of brothels nor an organised system of bar-based sex trade. In fact, third-party involvement in the organisation of the trade is rare. Most CSWs operate independently (see Chapter 4). In Kenya's coastal region, however, beach operators (commonly known as 'beach boys') play a critical role in the operations of CSWs (see also Chapter 8).

The role played by the beach operators/beach boys in Kenya's sex industry, especially on the coast, warrants a brief discussion of how this informal sector of the national tourism industry operates. As mentioned by earlier commentators on the subject, a Kenyan beach boy is not a boy at all, but rather a young member of the host community, who, as the name suggests, initiates his trade on the beaches frequented by tourists (see, for example, Kenya 1998; PTLC 1998; Dahles and Bras 1999; Kibicho 2003).

Three categories of beach operators are recognised and regulated by the Kenyan government (Kenya 1998; PTLC 1998): curio-dealers (who sell curios to tourists on the beaches), local boat operators (who take tourists for deep-sea safaris) and safari sellers (who sell various safari packages to tourists on the beaches) (Kibicho 2003). The rest of operators on the beaches in Kenya are referred to as 'beach boys'. They include CSWs, massage parlour operators, moneychangers, self-styled interpreters and guides, drug traffickers and food hawkers. Their income is normally irregular, and they may go without business for days. To cushion themselves from this uncertainty, they spread themselves thinly over a wide range of deals rather than plunging deeply into any single one (Kibicho 2003; Dahles 1999). Other risk[9] reduction or avoidance strategies include working long hours, seeking supplementary income, trading in small quantities, imitating products and services offered by others, assuming many business costs themselves, and being reluctant to grow (see also PTLC 1998).

Beach operators are involved in the provision of conventional tourist products and/or services only in marginal ways. In general, there are two market niches for them in Kenya's coastal region:

1. provision of tourist products/services which lie outside the commercial interests of the large-scale firms; under this market category, they are involved in handicraft production and local transport (taxi) services;
2. provision of services that complement tourist services controlled by the dominant, large-scale multinational firms; thus, they are involved in localised safaris to deep-sea areas and/or to nearby inland PAs, town guiding and retailing of cheap souvenirs. (see Kibicho 2006c, 2006d)

---

9 In this context, risk is defined as the unintended consequence(s) of a rational action.

Rasta Culture as a Sex Selling Point

It is important to open this section by noting that the Rastafarian movement is a culture of resistance that traces its roots to Africa. It cherishes its black African heritage, and commits itself to a natural way of life. The principal characteristic of the Rastafarian is growing dreadlocks that symbolise the strength of the lion and signify pride in African heritage. Rastafarians or Rastamen grow their hair long and exhibit a leisurely lifestyle dedicated to reggae music (Pruitt and LaFont 1995). The movement repudiates material accumulation and participation in the system of exploitative lifestyles (Pruitt and LaFont 1995; Dahles 1999).

To entice clients, Kenya's beach operators have adopted Rastafarian attire and behaviour. Rastafarian attire is also used to signal the wearers' availability as escorts for female foreign tourists. All the female (sex) tourists interviewed (see the section 'Female Sex Tourists above), for instance, reported that they were aware of this signal. One female (sex) tourist noted:

> you need not to go to school to be able to differentiate between what is on or for sale and what is not. Remember, the old adage ... if it looks like a duck, quacks like a duck and walks like a duck – then we professionals [female sex tourists] know it is a duck – a MCSW [she ends with uncontrollable laughter].

Thus, the Rastamen's adopted patterns of dress and behaviour allow them the access to tourist space and time (Figure 9.2). They indeed represent a major attraction for female tourists from the Western world. As Dahles (1999: 188) puts it: 'for most foreign women ideal(ised) features of masculinity are associated with the Rastafarian'. The Rastaman is thus constructed as the exotic Other, more passionate, more natural, more emotional, and most important, sexually more tempting than Western men. In fact, a study in Jamaica by Pruitt and LaFont (1995) found that those local men with dreadlocks received substantially more attention from foreign women than did their counterparts without locks.

To some of these local men, wearing dreadlocks also signifies sexual prowess, marijuana, naturalness, erotic-exoticness, and liberation from all forms of oppression. Paradoxically, the ultra-macho posturing of these beach boys is only momentary, as they quickly occupy the feminine subject position in their dealings with Western female tourists. It should be noted that 'it is sickeningly un-macho for a man in a traditional African patriarchal culture to giggle and flirt, and to walk down the street holding hands with a woman' (Mwangi 1995: 2). In relation to the foregoing, Backwesegha (1982: 17) is of the opinion that 'all the things a beach boy must do to get a Western woman to go with him demeans his masculinity in local cultural terms'. To him, it seems, to agree to follow a female (sex) tourist, bearing in mind that she is older and more successful, at least financially, means losing control and weakening towards local social norms. In other words a local man–female sex tourist relationship reverses the traditional African male–female model, as local young men (beach boys) prostitute themselves to dominating

Western women who are in most cases considerably older than themselves (see also Cohen 1982, 1988b, 1989; Dahles 1999; Kibicho 2005b). It should be reiterated that in the traditional African societies, sexes almost never socialised in public, or at least never showed their affection openly (Mungai 1998). The display of romantic involvement was only approved in the 'appropriate' place – the bedroom (see, for example, Backwesegha 1982; Mwakisha 1995; Mwangi 1995; Ndune 1996; Mungai 1998; Sindiga 1999; Afrol 2003; Kithaka 2004).

In an attempt to mask the fact that they are performing feminine roles, beach boys-cum-MCSWs often stress and assert their masculinity to their audience. This is normally done through an extraordinary emphasis on their sexual prowess, on their ability to manipulate and control their Western women lovers, or by simply talking too much and/or too loudly in public. In other instances, local beach boys opt for consumption of copious quantities of alcohol (Figure 9.2). To them, alcohol intake is of significance as a means of encouraging disinhibition, since their expectation of intoxification is that it allows 'emancipation' from social rules, increases sexual confidence, and frees the individual from responsibility for his action. In general, therefore, the MCSWs counterbalance their powerlessness *vis-à-vis* Western female tourists-turned-'girlfriends' and their local community by developing a virile and a strong, self-confident self-image. Thus, one may say that the MCSW–female tourist relationship serves a dual purpose – it provides the sex worker with an opportunity to escape from his hopeless financial situation, while at the same time creating a situation in which the concerned men can at least pretend to control the Western female visitors who are in a structurally more powerful position. It thus creates a superficial impression, at least from a theoretical viewpoint, of these men manipulating a world that appears designed to suppress them.

There is definitely a destructive potential inherent in these relationships. This assertion is based on the fact that what is a pleasant and refreshing interlude to a visitor could lead (both in the short and long term) to the destruction of at least one of the foundations of the local social set-up: that of ordering social life according to gender, age and generation gaps. When a woman assumes the traditional male role of provider and the man becomes dependent on her, for instance, this would seem unacceptable, as it contradicts basic values in the African social structure concerning appropriate gender-role behaviour (Ndune 1996; Mungai 1998; Sindiga 1999). It violates social standards of normalcy.

However, these effects should not be oversimplified, as this would result in a superficial understanding of the subject. They should be understood within the context of both established local gender relations and new economic and cultural opportunities offered by the emerging tourist culture in tourism destinations. For example, if changes in these gender-role behaviours are based on a man who purposefully seeks support from a woman 'as a route to financial and/or material success, it is then contradictory to emphasise on what constitutes the masculine role' (Dahles 1999: 183). This understanding should therefore be underpinned by the fact that all tourism destinations are social spaces where disparate cultures meet, clash and grapple with one another, often in asymmetrical relations of domination and subordination (Dahles 1999; Wels 2000). The concept of social space or contact zone invokes the spatial and temporal co-presence of people previously separated by geographic and historical disjunctions, whose paths now intersect (see also Fanon 1966; Mathews 1978; Cohen 1993; Dahles 1999; Kempadoo 1999).

These reversed gender relations can be understood fully from the perspective of structurally unequal power relations between tourist-generating and tourist-receiving countries which sneakily restore the rules of colonial sexual decorum (Figure 9.2) (see Britton 1982; Cazes 1992; Hall 1994; Cuvelier 1998). The phenomenon in which (older) women travel to tourism destinations with the intention of enjoying sexual encounters with the financially poor local (young) men is a textbook case representing the exploitative nature of much of the tourism business between the developed and less-developed countries (see also Chapter 3) (Britton 1982; Burns and Holden 1995; Brenner 2005).

Conclusion

Sex workers' extractive modes of earning a living and tourists' equally creative ways of gaining satisfaction represent phases of contrasting operational complexes, but one depends on the other. On a different note, however, one may say that tourism does not create (child) sex trade, but it provides a means for easy access to vulnerable children. It provides an outlet for individuals with a predisposition for child sex that cannot be satisfied as easily in their home countries.

Moreover, the burgeoning tourism industry in Kenya has opened up small-scale business opportunities in the sex trade as the indigenous (sex) industry adjusts to cater for a mass market. The changing role of women, and economic hardships experienced by Kenyan men, appear to pose a considerable threat to the traditional male identity. Subsequently, the lost paradise of natural sensuality is embodied by the youthful Kenyan males (beach boys) who are ready to please women (tourists) of all ages and origins. This mythology runs parallel to the desexualisation of women. The attraction of Kenya's male beach boys to sex liaisons with female (sex) tourists must accordingly be perceived in relation to the loss, within the normal socio-economic environments for the conventional male role of financial supporter. A combination of these socio-economic changes has transformed Kenya's key tourism destination areas into havens for both domestic and international (sex) tourists. To understand the manifestation of the tourism-related sex trade in Kenya, Chapter 6 discusses sex tourism in one of the country's major tourism destinations – the coastal region.

Chapter 6
Socio-economic Base of Sex Tourism in Kenya's Coastal Region

Introduction

Faced with economic hardships, some Kenyans have devised various means of survival, including selling sex (see Chapters 1, 3 and 9). This chapter examines the relationship between tourism and sex trade in Kenya's coastal region. With its 800-kilometre stretch of pristine tropical beaches, this region attracts the majority of Kenya's one million-plus tourists every year (see Chapter 3) (Bachmann 1988; Sinclair 1990; Dieke 1991; Sindiga 1999; Akama 1999; Kenya 2004, 2006). Thus, Kenya's tourism industry is heavily oriented towards its coastal regions. Consequently, beach tourism contributes over 60 per cent of the national tourism activities (see 'Setting the Scene') (Kibicho 2003).

Seven resorts – Bamburi-Kisauni, Diani-Ukunda, Malindi, Mombasa Island, Mtwapa, Shimoni and Watamu – were chosen as study units (see Figure 6.1). This selection was based on the fact that these resorts are the main tourist concentration areas in the region. As noted earlier, a questionnaire survey was then conducted among CSWs in the seven tourist areas during January–July 2002. Age, origin, areas of operation, reasons for being in the trade and risks involved in the profession were some of the principal topics of investigation. This chapter concentrates on factors inducing CSWs into sex trade, while Chapter 7 focuses on the challenges associated with the sex profession in Kenya's coastal region.

This chapter draws upon empirical evidence and attempts to synthesise CSWs' views and the arguments used by previous authors working from different theoretical perspectives to arrive at an understanding of the relationship between tourism and the sex trade in the Kenya's coastal region. Is it possible to put any empirical content into this theory? Are the concepts defined earlier real and measurable? Thus, the following pages present the findings of a study designed to test whether the theory works in the real world (see also Kibicho 2005b). Globally, it aims to determine why Kenyans engage themselves in sex trade. Combined with Chapter 7, therefore, the chapter bridges the discussion in the past five chapters and the final two chapters, which examine tourism-related sex trade in the Malindi Area (Figure 6.1) to crystallise our arguments.

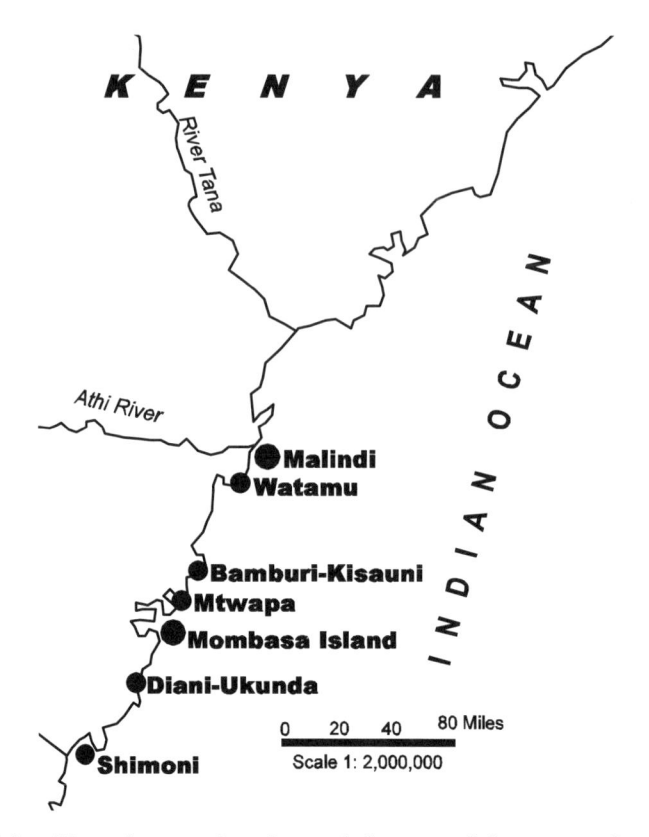

Figure 6.1 Kenya's coastal region and the area of the case study

Tourism's Seasonality and Sex Trade

Kenya's coastal tourism, like its national tourism, is highly seasonal, marked by two main seasons – peak season being November–March and off-peak season April–July, with mid-season spanning August–October (see also Chapter 3).

The off-peak and mid-seasons have always been low-business tourism periods in Kenya's coastal region. This forces hoteliers and other tourism operators to lay off or send some of their staff on unpaid compulsory leave. In some instances where the hotels are closed due to low bookings, all the workers are sent on unpaid leave until bookings improve (PTLC 1998; Kenya 2000). According to Thoya (1998), a total of 50,000 people working in various tourism establishments had been declared redundant by the end of 1997 after many international tourists cancelled their visits to the region. These cancellations were due to ethnic clashes which erupted in the month of August the same year. Table 6.1 illustrates that some of the hotels in the region closed down in the period September 1997–March 1998.

Table 6.1   Hotels closed September 1997–March 1998

Region	Hotel	Star rating
Mombasa	Shanzu Village	3
	Giriama Beach	3
	Bahari Beach	4
	Oyster Beach	3
	Malaika	2
Malindi/Watamu	White Elephant	2
	Jambo Club	3
	African Dream Village	4
	Casauarina Villa	3
	King Fisher	2
	Bougan Village	3
	11 Villago Graooss	2
	Karibuni Hotel	2
	Mayungu Holiday Resorts	3
	Blue Marlin	2
	Indian Ocean Lodge	2
	Coconut Village	2
	Baracuda Inn	4
	Temple Point	4
South Coast	Trade Winds	2
	Two Fishes	4
	Diani Sea Lodge	3
	Neptune Paradise Hotel	3
	Nomad Beach	3
	Club Green Oasis	3

Source: Kibicho (2005b: 261).

In most cases, those workers who are laid off do not go home and wait for the tourism season to improve, but rather seek temporary engagement elsewhere. As noted earlier, the ever-shrinking labour market would not meet the sudden demand for job opportunities, and absorbs only a few. Some of the unlucky men and women who fail to get an alternative job try their luck in the informal sector, including sex trade.

In some instances, young girls indulge themselves in casual sex for pleasure, though not necessarily with tourists. Sometimes they get pregnant in the process. If such a girl is dumped by her 'boyfriend', she faces a lot of difficulties in her early life, as abortion is both illegal and against local traditions. Due to limited or no support from her parents, she may be tempted to look for alternative sources of money (see Jennifer's story, discussed later in this chapter). But even without the prospect of pregnancy, the girl may be introduced by her peers to the 'coastal life'. Here, as Ndune (1996: 6) found, she will hope to get herself 'a white tourist who might turn her into a millionaire overnight'. He further notes that 'the young girl knows her age advantage would outsmart the older CSWs and thus gives herself the label of *manyanga* (school/young girl) and advertises herself as such from the outset'.

However, there are other cases of child sex workers who have been sent out by their parents to beg for money (see also Chapter 5). Initially, the intentions may be sincere, because they may have been driven out by hunger. Unfortunately, among the many tourists there are some who will take advantage of such unsuspecting girls (see also Chapter 2). This explains why there were some very young CSWs reported in Kenya's coastal region during the study period. For example, the youngest respondent in the current study was 10 years old, with no formal education (Chapter 5). This is a good example of child labour which violates one of the key tenets of sustainable tourism. Once introduced into the practice, they will be hooked on it until they graduate to being full-time CSWs (Cohen 1993; Burton 1995; Ndune 1996).

For a clear understanding, also consider the case of the Philippines, where most young women were initially employed in the family business that catered for tourists and did not need to turn to sex trade for survival. However, a few did fraternise with male tourists, hoping to get married, which would assure them upward mobility outside the Philippines (Cohen 1988b; Smith and Eadington 1992; Kempadoo 1999). Sometimes, Kenya's CSWs, with their dreams of a stress-free life in Western countries as opposed to their poverty-stricken backgrounds characterised by a daily struggle for a meal, get married to white tourists (Ndune 1996). It is worth noting that while at her new-found home, the woman may suffer from loneliness, and sometimes she may be forced into sex trade (De Kadt 1979; Ackermann and Filter 1994).

Moreover, in some cases, 'women have been tricked into marrying foreigners, sometimes with the help of illegal marriage brokers, only to end up working in brothels abroad' (Leheny 1995: 2). In Germany, for example, there is a scheme which pays German homosexuals to marry foreign women and bring them back in return for a large fee from a pimp (Graburn 1983; Phillip and Dann 1998; Ryan and Hall 2001). However, in some instances a relationship with a foreigner has proven to be a successful strategy for many young women who seek opportunities and prosperity unavailable in local society (Cohen 1993). Such cases serve as examples to those who have not made it – a symbolic affirmation that success is possible, thus they should keep on trying. Towards this end, it would be unrealistic to imagine that tourism development in Kenya's coastal region has no responsibility for the ever-expanding sex industry in the region, as it directly creates potential or new entrants to the profession. This happens because the local residents recognise the unmet needs and desires of the tourists. They then apply themselves to satisfying these needs, including sex-related ones.

Tourism development brings about the creation of new settlements, especially in the emerging tourism resorts (Holloway 1994; Burns and Holden 1995; Brown and Hall 2000). This happens as people from other areas are attracted by the tourism growth to come and work in the industry. In Florida, for example, Greater Orlando's population increased in the decade 1970–80 by 30 per cent compared to other major cities such as Miami (4 per cent) and Jacksonville (7 per cent). This was attributed to the development of Disney World and other tourist attractions (Holloway 1994). Kenya's coastal region (Figure 6.1) is experiencing a similar population expansion as Kenyans from upcountry migrate to the region in greater numbers in the hope of working in the tourism industry (see Chapter 3) (Kibicho 2003). On arrival, they are met by an unstable, and indeed reducing, employment structure, resulting from the depressed labour market, not only in the tourism industry, but also in the whole coastal economy. Confronted by this situation, many women opt for self-employment in the informal sector (which can absorb an unlimited number of operators) as curio vendors, safari sellers, massage parlour operators (on the beaches), town guides, hawkers, discothèque staff or bartenders. Associated with these kinds of jobs are limited earnings insufficient for their upkeep, not to mention their immediate dependants. With relatively higher income and easy entry into sex trade, as there exist no minimum entry requirements and no pre-set regulations on the trade, a number of women join the profession as an exit route from this predicament (see also Chapter 8).

As mentioned earlier, however, some commentators note that sex trade in the (traditional) African context is a social problem (Sindiga 1999; Ndune 1996). They underscore the need for a detailed treatment of the current subject in order to single out the 'malfunctioning' aspects of the socialisation process. Towards this end, however, one should keep in mind the fact that to define a socialisation process is to describe the series of social changes that the process involves, and most importantly, the effects that these changes exert upon the attitudes and values of all the persons concerned. It thus includes a complete description of all the social problems that are created by the (social) process. In any case, the possible solution of a social problem becomes relevant only when the disturbances in the attitudes and values of the persons affected are full and reliable (see, for example, Kelly 1988). The next section therefore addresses some of the root causes resulting in the expansion of the noted 'social problem' – sex trade – in Kenya's coastal region, and in fact in the rest of the country. The chapter thus goes beyond the fallacies of most of the previous commentators on the linkage between sex trade and tourism.

Reasons for Being on the Game

The routes into sex trade are varied (Cohen 1982; Caplan 1984; Kempadoo 1999; Brown 2000; Agrusa 2003; AVERT.ORG 2004; Kibicho 2004b). Some Kenyans make independent lifestyle choices due to the realities of economic needs in an economic climate of recession, inadequate benefits, unemployment and increasing debt. Others venture into trading in sex as a personal choice. For example, Julie (fictitious name), a CSW in Mtwapa (Figure 6.1), reported:

> There are many reasons why one goes into sex trade .... I went into it by choice and if I decide to quit, this might happen this year or next year, but I cannot see it happen because I am not ready to leave my good clients. I am not prepared to give up these 'warm' and 'lovely' tourist dollars. Of course, I am not in the [sex] trade because of low self-esteem. I have a high self-esteem as I have other skills. I am a trained primary school teacher, and I do not think that every sex worker has low self-esteem. Once you start stigmatising sex trade, CSWs start having low self-esteem. If sex workers believe in themselves, they can decide for themselves ... go back to college or work in a different sector of the economy. How many women prostitute themselves in relationships they do not want to be in, but stay in a marriage for financial gains? If it were not for financial gain how many women would walk out of these relationships? Surely, is this not prostitution? To me, sex workers do not sell their sexuality any more than does the married women. We [CSWs] only do it more openly and less hypocritically.

Amanda (not her real name), another CSW in Mombasa (Figure 6.1), was even more unequivocal, and indeed philosophical. She noted:

> I got to a point in my life where I was re-examining why I was doing what I was doing, why I was making the kind of choices I was making. Ultimately, I decided that the only person I could really trust as a barometer was myself. If I was going to succeed or fail in life, I would fail or succeed on my own instincts rather than try to pander to some 'imaginary audience' (public). I took a personal decision to be a CSW.

Men and women sometimes drift into the sex trade through association with friends already working in the sex industry. This can be referred to as entry into sex trade through peer association and pressure. An example is Sarah (not her real name), who became a CSW at the age of 19 through her friendship with other girls. She said:

> I had two girlfriends who used to work as CSWs. We used to go to nightclubs to have fun together, then I would wait for them to serve their numerous clients. They used to buy for me some beauty products in return. After some time, I felt the need to be closer to them; I wanted to be associated with them and to be able to exercise the group's values. The easiest way for me was to be like them since they could not be like me. I mean, they could not quit sex trade profession. This is how I entered the sex industry. I just wanted to belong (somewhere). It had nothing to do with money. I went in as a mixed up little kid. I just followed the waves .... I did not make a personal decision. My friends, of whom one is now

dead of cancer often associated with the virus that causes AIDS, made them for me …. I am ashamed to think about this ….

Overall, although many CSWs enter sex trade as a way out of relative poverty, a number of them are influenced by the local conditions – the existence of a number of women working as CSWs and the high levels of unemployment (see Figure 3.4). In other words, an option not thought of before presents itself, and a decision to try trading in sex is made. Other reasons for entry into sex trade include coercion by pimps, and need for money to finance drug habits, while others have to finance the drug habits of their partners. Others have entered sex trade simply to supplement their salaries.

Materially, therefore, sex trade is a response to unsatisfied needs, financial hardship and poverty in its general manifestations. Thus, trading in sex is generally a last resort when faced with poverty (see Chapters 1 and 9). We cannot therefore examine sex trade in Africa without looking at the socio-economic contexts which give rise to the trade. Many CSWs enter sex trade as a side job, have low social status and are low-paid. For instance, the main reasons for being in the profession, as identified by CSWs in Kenya's coastal region, are unemployment, family problems and loosening of the traditional African bonds (see Table 6.2). The remainder of this chapter discusses them in turn.

Table 6.2   Factors leading Kenyans into the sex business

Reason for entering sex trade	%
Unemployment	59
Part-time employment	19
Family problems	8
Prestige	6
Pleasure	5
Adventure	3

Unemployment Supports Expansion of Sex Business

Fifty-nine percent of the total respondents ($n = 183$) gave unemployment as the key reason driving them into sex trade. Sixty-eight per cent of the sex workers in this category have attained at least a secondary/high school level of education, with the majority aged 20–30. In spite of this relatively high level of education, these CSWs are unable to obtain alternative forms of employment. This can be attributed to high rates of unemployment in the country at large (see Chapter 3). The national unemployment levels are estimated to be 35–50 per cent (Kibicho 2007). However, it should be noted that the Kenyan government's official unemployment estimate is 35–40 per cent, while estimates by private and international organisations are 40–50 per cent (Kenya 2006; Kibicho 2007).

All the CSWs who identified unemployment as the key factor leading them into sex trade serve all types of customers, especially during off-peak tourism seasons, when tourism-related activities are minimal. For them, the client's background is irrelevant, what matters is the willingness and capacity of the customer to pay for the sexual services required. In most cases, as with coffee-shop girls in Thailand, the CSWs try to maintain liaisons with their foreign friends by mail (see also Chapters 8 and 9) (Cohen 1982, 1988a; AVERT.ORG 2004). The CSW concerned benefits from such liaison maintenance as in most cases, he or she receives uninterrupted financial assistance from the friend. It also ensures that the CSW has a ready market for his or her sexual services any time the friend revisits the region.

Most of these local 'boys' would prefer 'girls', female sex tourists, of their age (source: Field notes). However, they seem to understand that older women are often in a more secure economic position and thus have better purchasing power. This is simply why they are highly attractive sexual partners, as their economic status promises the local men a ticket to a better life (see the conclusion by Greg cited later in this chapter). In consequence, it is a common feature of Kenya's coastal region to find female tourists in their forties and older being courted by 'boys' in their early twenties. Similarly, female CSWs prefer older male sex tourists, but for a different reason. According to Julie, the CSW quoted earlier in this chapter:

> older men are sexually mature. They are loyal to their ladies [CSWs], remaining with the same girls [CSWs] whenever they revisit the destination. This is unlike their youthful counterparts who are always on the move, trying to sleep with as many local women [CSWs] as possible during their short stay at the coast. One would be excused for thinking that they are in a contest. We [CSWs] rarely have enduring relationships with them ....

Ninety-eight per cent of the respondents in this category are willing to quit the sex business if they can find a better alternative. Eighty-seven per cent of them would prefer to be assisted by the government to raise capital to start small-scale tourism-related projects like curio shops, tour guiding, taxi operations or boat-safari operations. This is a challenge to Kenya's government to come up with a micro-enterprise credit facility in order to absorb these large numbers of CSWs into alternative sectors of the tourism industry. The remaining percentage would opt to work in various sectors of the tourism industry – the hotel sector (6 per cent), tour operations (4 per cent), and environment conservation agencies (3 per cent). A clear contradiction, however, appears with the 2 per cent of the sex workers in this group who are unwilling to leave the sex profession. When asked about the possibility of leaving the trade, one respondent reported:

Let us stop pretending and being hypocritical .... I am here [in the sex trade] to stay. Many girls have come and left me in this trade. I am in love with my profession [sex trade] .... I like it. Of course, I know many people will think that I am insane, but to me that is immaterial. That is the reality – my reality. In any case, you can remove a girl [CSW] from prostitution, but you cannot remove the tendencies associated with prostitution from her. In other words, once a whore always a whore.

From a general point of view, this category of CSWs falls withing poverty sex trade in the three-tier model of sex trade (see Figure 1.1).

This high number of CSWs willing to quit the sex business necessitates a brief examination on how this enormous reservoir of potential human energy can be tapped (see Chapter 1). Such a strategy comprises three components: opportunity, empowerment and security. Opportunity means CSWs must have access to economic opportunities of which they can take advantage to change their destiny. This argument is based on the fact that a charity-like approach could make CSWs overly dependent on donations and lose their motivation to improve their lives by themselves. Therefore, emphasis should be put on the income capacity-building of the sex workers, in which economic opportunity plays a crucial, incubating role. The second component, empowerment, should aim at enhancing the capacity of the CSWs to influence socio-political institutions, thus strengthening their participation in political processes and local decision-making. This component should highlight the removal of barriers that work against CSWs, and instead build their assets to enable them to pursue and benefit from economic activities. However, simply expanding opportunity and empowerment are insufficient; to consolidate the results of tapping CSWs' potential, a social security system should be established. This component, security, should aim at reducing the vulnerability of CSWs to various risks such as ill health and economic shocks. In brief, these three components, each from a distinct perspective, manifest three requisite but supplementary approaches to assisting local people driven into sex trade by lack of job opportunities. However, to achieve the desired effect, all three components should be strengthened concurrently.

Part-time Employment to Supplement Income

The primary motive for going into sex trade for 19 per cent of the sex workers in Kenya's coastal region is to find part-work to supplement the small amounts they earn in their normal occupations. In other words, this reason is directly linked to the previous one, unemployment, or in some cases under-employment. Fifty-six per cent of them work during the day as secretaries in various establishments, with half of them working in tourist hotels. They have a tertiary level of education. They do not discriminate against their clients, especially during the off-peak tourism seasons. However, when the tourism industry is doing well, they only entertain tourists (both international and local) aged 40 and above. All the CSWs in this category would be willing to leave the sex profession if their pay packages were improved. Khadija (not her real name), a female CSW in Diani (Figure 6.1), observed:

> I work as a secretary in a three-star hotel in Diani Beach [Figure 6.1]. I earn a monthly net salary of K£345 [see Table 9.3 for the exchange rate]. This money is not enough to take care of my two children and my personal needs. This is without mentioning those of my extended family. To meet all these financial obligations, I had to look for a source of side income .... Due to the easy entry into sex trade, and of course its flexibility in relation to my working schedule, I chose it as my side job.

Khadija's case appears more a question of the need for survival than consumerism. This is quite similar to Thompson and Harred's (1992) observation that some topless and bottomless dancers chose sex trade to support their studies and their family. In many cases, the money they can earn during limited and convenient working hours is the major attraction for entering the sex profession.

Maggie (not her real name), another CSW who worked as a front office receptionist in a four-star hotel in Mombasa, noted:

> I work hard from Monday to Friday and sometimes Saturdays. I socialise with my clients [commercial sexual service consumers] over the weekend or occasionally in the evening after office hours. Of course, some of my colleagues are aware of my double life, but they can do nothing about it. It is my life and it is not affecting my work in the hotel. One career does not prevent me from pursuing the other; I even think I am one of the best in the two careers – receptionist and sex trade.

When asked how and why she decided to enter the profession, she gave the following account:

> It was early last year [2001]. I went with my lady friends to a discothèque in town [Mombasa]. Two hours later, I was introduced to a well-built male Israeli who was willing to buy me drinks. At this time, my [two] friends were paired up with guys and were having fun. After some hours of beer drinking, my new 'friend' wanted to leave. He invited me to accompany him. I turned down his request. He then offered to pay me US$300, if I spent the remaining part of the night with him. I was shocked! I demanded that he give the money [in full] to one of my friends before we leave. He obliged. Then I started wondering, 'What exactly is he going to do with me? What kind of sexual service is worth this kind of money?'

In short, this was the first time I received money for sex. I needed the money badly, but not sex. After this, I realised how easy it is to make a quick buck doing what I enjoy most. It soon became a habit .... Today, sex is like a drug to me, I take it when I want to feel high. Yeah, I get satisfied (both sexually and psychologically) and at the same time, I get richer. It is killing two birds with one stone – my precious 'asset'! [She laughs loudly, pointing below her abdomen.]

Maggie reports that on average she earns about US$1,060 and US$ 850 per month from her sexual services during the tourism peak and off-peak seasons respectively, while she is paid US$146 a month by her 'formal' employer. She notes that she works in the hotel to keep herself busy during the weekdays, when Mombasa's sex industry is relatively inactive. In her own words, she prefers to treat the work she does in the hotel as a side job. – A side job to sex work! This is contrary to many other CSWs in Kenya, and probably in the world, who work in the sex industry in order to supplement their earnings from jobs in the formal sectors of the economy (see, for example, Khadija's story at the beginning of this section). Overall, this group of CSWs falls within tourism sex trade in the three-tier spectrum of sex trade (Figure 1.1).

Family Problems Induce Some Kenyans into the Sex Industry

Of the total number of sex workers ($n = 183$), 8 per cent were induced into the profession by family problems. Some had been thrown out of their parental homes after having babies out of wedlock. Others, 10 per cent of respondents in this category, were victims of broken marriages. It is worth mentioning that while migration within or outside the country is reshaping the Kenyan family, it has in no way diminished the importance of, or the desire to be, 'family' (Ndune 1996; Mungai 1998). Strong kin linkages are derived from a combination of the cultural importance of effective ties, bilateral kinship and absence of institutionalised social security. As a consequence, marriage and childbirth are integral elements of a Kenyan, and indeed an African, family. Thus, Kenyan men and women are culturally expected to marry and have children regardless of their socio-economic class. This said, the possibility that a woman, in particular, might 'voluntarily' choose to perpetuate her single state rarely enters the Kenyan mind (Backwesegha 1982; Ndune 1996; Mungai 1998). In fact, in many African societies, women are not expected to leave their matrimonial home if they are battered by their husbands. The beating of women is widespread, even in the developed world, yet in many instances it is explained in individualised or situational terms (see, for example, Hanmer et al. 1989). Many societies profess to deplore wife battering, yet their deep-seated attitudes regarding it appear to be ambivalent (McLeod 1982; Hanmer et al. 1989; Hobson 1990; Mungai 1998). Blame for battering, for example, is frequently placed on the battered woman. Either she 'brought it on herself' in some specific way, or else she is held responsible for getting herself into such a situation (see, for example, Backwesegha 1982; Mungai 1998). If the woman is dependent economically and/or emotionally and her social standing depends on the man who is doing the battering, then the battered woman often find herself unable to leave. However, some Kenyan women do leave marriages where they have been beaten by their husbands.

In Shimoni (Figure 6.1), for example, Marion (not her real name), an uneducated and jobless woman with four children under the age of 10 to look after, was abandoned by her husband. Her only 'sin' was to complain about her regular beatings by him. This was aggravated by her questioning of his extra-marital sexual relationships. It begs noting that sexual double standards dating from British colonialism seem to have been remarkably persistent, encouraging restraint and fidelity among women, and licence among men to engage in casual sex and extra-marital affairs (see, for example, Mwakisha 1995; Sindiga 1995; Ndune 1996; Mungai 1998). Thereafter, Marion looked for any kind of job, with no success. Finally, she opted for the only readily available alternative – sex trade. She explained:

> The profession requires no certificate. The 'skills' are learned through interactions with peers [CSW–CSW] and clients [CSW–sex tourist] and successful accomplishments [see Figure 1.2]. It is not learned in a decontextualised manner. It is not learned by completing a series of lessons in commercially available programmes. You only need to 'dress appropriately', be at the right place at the right time and men will come running after you .... Yes, I do not like the sex job, but I like the money. Today, I can afford to buy food and clothes for my children ... But, I'll never forgive the father of my kids ....

Marion's story confirms that 'men are willing to pay more for sexual access than for almost all other forms of female labour' (Symanski 1981: 3). They may sacrifice/give all kind of resources to gain 'access' to women.

Jennifer (fictitious name), another CSW in Watamu (Figure 6.1) reported how she was 'forced' into sex trade by her parents after having a baby out of wedlock. She narrated:

> my life was fine before I became pregnant and eventually gave birth. You know I come from a family with a strong Christian background. My mother abused me regularly ... my father was even worse. He battered me on a daily basis. They hated my baby girl and thus gave me no money to buy her food and clothing. As if that was not enough, my boyfriend denied that he fathered my baby. At that time, I was only 18 years old. I had no source of income .... I explained all this to one of my lady friends who kindly introduced me into Malindi night life. Now, my daughter is four years old. I am able to buy her food, clothes and presents. But I pray that she does not take this profession when she grows up. I am always like, 'Okay, God, if there is an open door for me somewhere, do

not let me miss it.' This is what I always pray. In brief, I am here because of my parents, who pretend to be strong Christian believers. What is so unchristian in having a baby? [she asks amidst sobs].

Poverty Breeds Family Problems

Previous researchers have recognised the complexity of the context – social, cultural and individual – in which sex trade occurs (Harrell-Bond 1978; Backwesegha 1982; Cohen 1982; 1988a; Archvanitkul and Guest 1994; Brown 2000; ECPAT 2002; Agrusa 2003; Hall and Tucker 2004). This complexity is clear in Kenya, where most of the CSWs are affected by poverty and the related problems it breeds (see Chapters 1, 6, 8 and 9). In fact, in some regions, initial progress in dealing with poverty has been followed by continuing setbacks. These setbacks are all the more threatening for the ability to deal with the spread of poverty sex trade because they are occurring at a moment in history when traditional values, family organisation and economic structures are rapidly changing (Kibicho 2005b). For example, a major increase in non-marital births (especially to young, poor women) has moved many women and children into poverty (Mungai 1998; Gitonga and Anyangu 2008). The problem is aggravated by the fact that the fathers of many of these children do not take seriously their responsibility of providing for their offspring (see Marion's story discussed earlier). Further, the government does not enforce child support laws to ensure that the (absent) fathers meet their responsibilities. It should be noted that many poor youth are different from rich teens in that they see little or no hope in their futures. As a result, they often drift into pregnancy and then into parenthood. For most Kenyan teens, early childbearing immediately worsens their quality of life, and often leads to a number of negative consequences as far as their educational, economic and marital futures are concerned.

We now return to the issue of youth poverty, which is often the key factor for the social problems of Kenyan youth. Let us be clear from the start: no other change in Kenya's family structure has been more controversial than the rising number of non-marital births (Ndune 1996; Mungai 1998). According to African traditional values regarding family structure, non-marital births are seen as a rejection of the two-parent family (Mungai 1998). Of course, many Kenyan youth manage to navigate successfully through adolescence without experiencing (or causing) pregnancy, contracting STDs or bearing children out of wedlock. There are various explanations of how these teens move through this reproductive minefield of their sexual life. Some youth, for instance, are less sexually active, and in some cases even abstinent, while others are sexually active but take care to avoid pregnancy, or if they become pregnant, rely on abortion, albeit in a clandestine fashion. Paradoxically, Kenyan teenagers are today exposed to all kinds of messages about sex. Movies, music, radio and television tell them that sex is exciting, romantic and titillating, and pre-marital sex and cohabitation are familiar ways of life among the adults around them (Mwangi 1998). Almost nothing that they see or hear about sex informs them about 'safe sex' and the responsibilities of both partners, male and female, in sexual activity (Kibicho 2004b).

However, the increase in non-marital births by adolescent girls, mostly from poor families, brings to the fore negative aspects of blaming the girls concerned. It begs highlighting that in Kenya, youth birth rates are highest in households that have the greatest economic disadvantages (Kenya 2006; Gitonga and Anyangu 2008). As a consequence, one may say that poverty causes teenage childbearing. After childbirth, the teen mother faces more financial hardships which aggravate her already precarious status. In this case, teenage childbearing breeds poverty. That said, understanding the relationship between poverty and teenage pregnancy, comprehending the extent of the problem, interpreting its effects on African youth and children, and finding a sustainable solution are indeed highly challenging tasks. This explains why many commentators are unable to sort out the effects of early childbearing from the selective factors that induce teens to become youth parents (Afrol 2003). Anyhow, Kate (pseudonym), a CSW in Bamburi (Figure 6.1) reminds us:

> if you are really looking out for the best of human beings, you want to …vow to work for alleviation of poverty and figure out ways to avoid more poverty driven sex trade. That is how you can honour the CSWs' offspring.

As with the CSWs who identified unemployment as a motivator leading them to sex trade, 90 per cent of the respondents were ready to leave the trade if an alternative source of earnings were made available to them. Further, they would like to work in the tourism industry, especially within the coastal region. Using the three-tier categorisation (Figure 1.1), this group of CSWs falls within the poverty sex trade class, serving all customers willing to pay for their sexual services.

Sex Worker–tourist Relationship: An Association for Prestige?

Six per cent of the CSWs ($n = 183$) felt it prestigious to associate with white tourists, and since being in the sex trade is one of the sure ways to get such an opportunity, they find themselves in it. An interesting fact to underline here is that all those who gave this as a reason for being in the profession were from the coastal region, had comparatively low levels of education (primary level) and were in the 20–30 age group. Having dropped out of school at these early ages, they saw it as an achievement to have a white boyfriend. It can also be argued that due to their age and comparatively low levels of education, these sex workers were not exposed to the 'outside world' and were still in a colonial mindset. They saw a white man as a master, thus associating with him was prestigious.

However, there are no CSWs in Malindi Area (Figure 6.1) who are in the sex business for prestige reasons. This can be attributed to higher education levels in this tourism resort compared to other areas in Kenya's coastal region. Thus, as

Kibicho (2004b: 11) argues, 'these CSWs have sent-off colonial hangovers …. They no longer see a white man as a master and therefore associating with him is not necessarily prestigious.' Informal observation suggests, however, that at least some significant vestiges of this general pattern of the 'white master mentality' persist among some Kenyans with higher education. This tendency can only be explained by the self-fulfilling prophecy, the possibility that a false definition – if acted upon sufficiently – can, in effect, become true (Kibicho 2005c; Mensah 2005). Thus, the individual who has consistently been treated as inferior – and who also has been denied the opportunities to develop and demonstrate the capacities that would disprove this – may even come to see himself or herself as inferior. This possibility is obviously closely tied to Africans' 'servant status' during the colonial era. As a result, it is impossible to comprehend the intricacies of the modern 'white master mentality' without examining the links with colonialism in Africa.

As Mensah (2005) reminds us, many contemporary Africans' 'feelings of inferiority' are the consequences of colonial conquest and exploitation. Through colonialism, Western racial ideologies were deployed throughout the colonies, portraying whites as superior without any objective attempt to prove the superiority of other racial groups. White master ideologies were complementary to the colonial enterprise (see Chapter 3). Thus, racial doctrines were often invoked to justify master–servant relationships (Hall 1994, Wels 2000; Mensah 2005). As a consequence, relationships in Africa were structured around racial and ethnic differences. These differences translated into cultural superiority and ideological dominance (Wels 2000; Mensah 2005). Further, through discriminatory laws, bureaucratic procedures and military might, the colonial powers in Africa created an insurmountable divide between the coloniser and the colonised in all spheres of life. The force with which the 'superiority' of the white race and values was affirmed in the African colonies still evokes feelings of incompleteness among many indigenous Africans in post-independence Kenya. How else can one explain the view that by associating with a white tourist, a CSW feels prestigious?

Marriage into this category, referred to as 'open-ended prostitution' by Cohen (1982), is the overriding aspiration for all the CSWs willing to leave the trade once they get married to a white man. Other things being equal, sex workers who are in the sex business for prestige reasons fall within the tourism sex trade category (Figure 1.1).

Selling Sex for Pleasure

Five per cent of the sex workers ($n = 183$) in Kenya's coastal region, who seemed to be the elite of the sex profession, were in the trade for reasons of pleasure. They had comparatively high education levels, 64 per cent of them being university graduates. Their target clients were the local elites and tourists, both domestic and international, in the age bracket 20–31 years. They were from well-to-do (upper-class) backgrounds, and thus could easily enter other well-paying careers.

For them, money was not a motivator for entering sex trade. They were pleasure seekers. A good example is Amina (not her real name), a female CSW in Shimoni (Figure 6.1), who reported:

> I feel non-inhibited when having sex with strangers [tourists]. I am able to burst through barriers, such as upbringing, beliefs and religion to get the kind of sexual satisfaction I want. There are many inhibitors put in place by our archaic traditions to hold us back .... Of course, I love sex-dollars, but for me the most important thing is to enjoy sexual fantasies with strangers ....

In some instances, however, some of these (sexual) pleasure seekers became exclusively interested in money, and at least for a while, turned their emotions off completely, which can greatly be attributed to involvement in their peer group of other CSWs.

Their membership of informal sex trade-related associations like the MWA (discussed in detail in Chapter 9) is not based on the potential assistance given by the association, but rather a need to be 'closer' socially to other CSWs in their areas of operations. Rita (fictitious name), another CSW in Malindi (Figure 6.1), noted:

> being a member of the Association gives me an opportunity to interact with other lovely Kenyans who think and behave like me. It gives me a chance to learn many things from my like-minded sisters and brothers. It is the only organised forum where we can discuss our professional issues openly. Our Association also serves as an effective communication tool between the sex workers. It provides us with advice and moral support to ensure that the differences of opinion do not turn into major conflicts. Furthermore, working together makes us stronger and more powerful in relation to other un-organised sisters [CSWs] and the police ....

The grouping together of un-organised CSWs, sex workers who are not registered with the MWA, and the police confirms the high levels of rivalry between the two groups of CSWs in Malindi (see also Chapter 9).

Trading in Sex as a Form of Adventure

It should be noted that adventure as a reason for entering sex trade was added to the questionnaire by the CSWs during the interviews. However, a critical scrutiny shows that the reasons have more to do with economic needs than adventurism. Anyhow, for the sake of a discussion, we present it as a separate reason.

Three per cent of the respondents were in the profession for adventure purposes. They were excited at 'moving out' with strangers. Eighty-three per cent of this group came from upcountry. All of them enjoyed staying with international tourists for the rest of their stay in the region, and always accompanying them. This is a clear illustration that these CSWs were not only merely sexual objects, but also tourists' personal brokers. Sometimes they shifted their operational base from a rented room to the hotel where the tourist-turned-friend was staying. This was the group which occasionally married foreigners (Cohen 1993). None of the CSWs in this category was ready to quit the profession.

Greg (not his real name) was a good example of this kind of CSW in Kenya's coastal region. He was one of hundreds of young men working on Kenya's beaches – beach boys (see Chapter 5). He lived in a room with no electricity, no running water and no provision for refuse disposal. In contrast, the hotels and apartments that line the beach across the road are luxurious, with rooms costing as much as US$125 a night. Greg had a 'regular girlfriend', a 45-year-old Italian professional who came to see him four times a year. 'She's a good friend and she looks after me. She sends me money to help me out of financial difficulties, like when I am unable to pay my house rent.' Greg said he worked 'in the tourism industry', and would not admit that he was a beach boy or a CSW (see also Chapter 5):

> If I take a tourist out, and she wants to help me out as a 'friend', give me money and let me stay with her in the hotel, what's wrong with that? Of course I have sex with them, but that's because I am not gay – I like women.

He further observed:

> while some beach boys may be content to have their meals paid for, the ultimate goal is marrying and migrating, preferably to Europe or the US. This gives you a 'ticket' out of poverty. It gives you an opportunity for a new life and better days for your children. In any case, we are all life-adventurous [he concludes with a serious expression].

This assertion by Greg confirms our earlier observation that these CSWs were in the sex trade for economic-related reasons, and not for adventure. Furthermore, because it is never certain whether a relationship will meet the CSW's expectations, these sex workers spread the risk of an eventual break-up by being engaged in several relationships at the same time (see also Chapters 5 and 9). Certainly, this is not an adventure-oriented strategy.

In addition to being in the sex business for adventure reasons, some CSWs see it as just another line of work. It is a form of sexual liberation. Faith (fictitious name), a female CSW in Mtwapa, observed:

> Despite being an adventurer, I am driven by the desire to live a comfortable life, be as liberal and free spirited as I can. Yeah, some of my friends say that I am a social rebel, but to me, I am simply a woman who goes with what her spirit wants. I go for what I want, and do what pleases me first; other people or things come second …. Why should I please others when it does not help me? I enjoy

uncommitted sex. The advantage I have is that I choose who to 'play' [have sex] with and I am paid for it. I want to live my life to the fullest.

These two groups of CSWs, who are in sex trade for pleasure and adventure reasons, fit in the high-class sex trade category in the three-tier system of sex trade detailed in Figure 1.1. Unsurprisingly, all of them were unwilling to leave the sex profession. A question for the moralists: What should be done with this category of CSWs?

Other Predisposing Factors for Sex Trade

African Woman and Children

As noted earlier in this chapter, an African woman is incomplete if she cannot conceive and is unable to raise a child. Marriages are often affected if a couple is unable to bear a child (Ndune 1996; Mungai 1998). This said, greater social stigma is attached to women who cannot have children. Thus, the need to bear children is not confined only to those who can feed, educate and take care of them. Even beggars in the streets will strive to have children without worrying about their upkeep (Backwesegha 1982; Ndune 1996). Such women 'work' towards getting a baby. Once they get it, they are confronted with the financial obligations which go with raising children. With no job opportunities on offer, they find sex trade an easy way of generating income for their needs and those of their newborn babies.
The foregoing is complicated by the subordinate status of women and girls in most African societies, which, together with the intransigent poverty, conspire to force them into difficult working situations (see also Chapter 1). Such situations expose women to sexual abuse, with attendant risks of HIV-AIDS infection (see Chapter 7). Even without contracting STDs, some of these women become pregnant in the process. They then find themselves in a similar situation to that discussed above, where they have an added financial responsibility for the upkeep of their babies. In such a situation, the institutionalised exploitation of women within African patriarchal societies has been extended, and indeed systematised, by the unequal power relationships that exist not only between genders and members of ethnic communities, but also between the tourist hosts and the advanced capitalist societies (see, for example, Britton 1982; Findley and Williams 1991; Hall 1994, 1996; Brown and Hall 2000; Hall and Tucker 2004).

Loosening of Traditional Bonds

In traditional Kenyan society, sex was only approved of within marriage, and there were constraints which adolescents were required to observe regarding sexual activities (Kibicho 2004a). However, young people have tended to ignore or reject these traditional values, replacing them with those of their own which usually

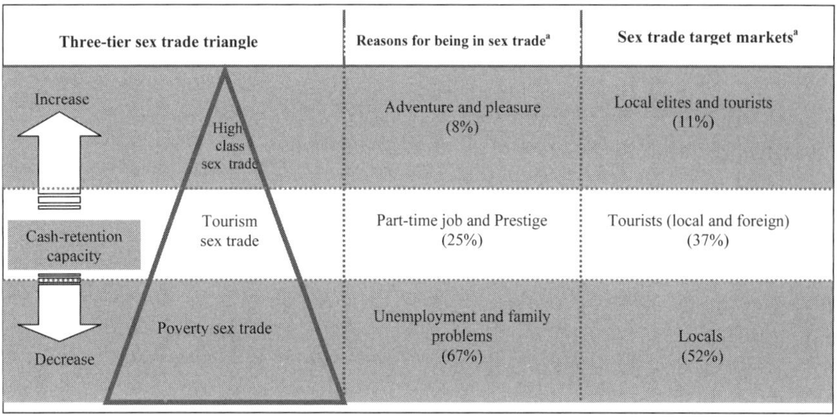

Socio-economic Base of Sex Tourism in Kenya's Coastal Region   139

Three-tier sex trade triangle	Reasons for being in sex trade[a]	Sex trade target markets[a]
High class sex trade	Adventure and pleasure (8%)	Local elites and tourists (11%)
Tourism sex trade	Part-time job and Prestige (25%)	Tourists (local and foreign) (37%)
Poverty sex trade	Unemployment and family problems (67%)	Locals (52%)

(Cash-retention capacity: Increase ↑ / Decrease ↓)

Figure 6.2 Categorisation of CSWs in Kenya's coastal region

Note: [a]Figures in parentheses are percentages in relation to the coastal region's sex business.

fail to live up to these ideals. For instance, in modern life, many young Kenyans have the belief that couples who intend to marry need to 'know' one another sexually before marriage (Mungai 1998). It is important to emphasise that these sexual activities take place at a stage when the players are incapable of assuming responsibility for any of the consequences that might occur, such as pregnancy, termination of education or contracting STDs.

The inherent conflicts with parents may force young girls out of their parental homes, thus raising their possibility of entering the sex industry (see Jennifer's case discussed earlier in this chapter). Moreover, the traditional control and influence of parents has been fading steadily in the face of many other contemporary forces for social change, such as education and technology, with their associated Western culture, the upsurge of tourism being a good example. This further explains the 5 per cent of the respondents who were in the sex trade because of family problems.

Based on the reasons for being in the sex trade and the CSWs' target markets, sex business in Kenya's coastal region can be summarised as in Figure 6.2.

According to Figure 6.2, 67 per cent of the CSWs in Kenya's coastal region fall within the poverty sex trade category. They have been driven into sex trade by lack of job opportunities (59 per cent) and family problems (8 per cent). They mainly target local inhabitants for their sexual services. Twenty-five per cent of the total sex workers in the region were either in the sex business to supplement their net earnings (19 per cent) or for prestige reasons (6 per cent) – they were excited to be associated with white tourists. Their target market was the tourism industry, and they therefore operated within the tourism sex trade in the three-tier sex trade spectrum (see also Figure 1.1). The remaining 8 per cent of the CSWs in the region worked within the high-class sex trade, as they had a highly specialised market, local elites and tourists, and their cash retention capacity was relatively high. They were in the sex profession for adventure or pleasure reasons.

Choosing Areas of Operations

The choice of areas of operation by CSWs in Kenya's coastal region depended on the sex workers' age. Those above 40 years of age chose areas with less competition from the young and outgoing women. Women in the 20–30 age bracket did not care about competition, but rather went for areas with high tourism activities, in this case Mombasa Island and Bamburi-Kisauni. As a result, sex trade in Kenya's coastal region was exclusively high in areas with high levels of tourism activity: Mombasa Island, Bamburi-Kisauni, Malindi, Mtwapa, Diani-Ukunda, Watamu and Shimoni (Figure 6.1).

The results show that Mombasa Island was the favourite for 30 per cent of the sex workers. The reason given for choosing the island as their favourite operation area was that there is always a ready market for sexual services. This is more so when a tourist cruise ship docks at the port of Mombasa. Male (cruise) tourists travelling without spouses are easily tempted to buy sexual services from CSWs at the visited destination (Backwesegha 1982; Cohen 1982). When a cruise ship arrives at the port there is a sudden creation of demand for sexual services to meet the erotic fantasies of men (see Chapter 4) (Cohen 1982, 1988b, 1993; Archavanitkul and Guest 1994).

Moreover, young women tend to move to urban centres, like Mombasa Island, with the aim of getting wage earners there, or even tourists, to purchase their services (Archavanitkul and Guest 1994). In such situations, the massive growth of sex trade in the island represents an attempt by the indigenous sex industry to meet the demands of the created mass market.

Small-scale Business: The Way Forward

Despite the fact that a few CSWs in Kenya's coastal region did not know exactly what kind of job they would prefer, 89 per cent of them dreamt of another job in the future. They were of the opinion that any other job would represent progress compared to sex trade.

Eighty-three per cent of the CSWs knew the kind of job they wanted, aspiring to careers in, or linked to, tourism. An example of a former CSW who owns a curio shop in Malindi downtown core and a massage parlour in the beach was frequently referred to during interviews with CSWs: 'She has four employees, all locals, in her two shops .... She is now living a decent life. Earning clean money. I admire her. She is our role model ...,' a CSW reported. Apart from massage, the said parlour also offers other beauty services, like pedicure/manicure, ear piercing and hair braiding. These services are often combined with the sale of small souvenirs such as woodcarvings, soapstone carvings, batiks, *lesho* and postcards. This is, of course, without mentioning commercial sexual services. With regards to this last service, the massage parlour owner complained of many sting operations by the police which were costing her a fortune in terms of money paid for bribery (see Chapter 4).

If the government is to help these people, then priority needs to be given to training programmes on management and marketing. This kind of training will ensure that CSWs-turned-small business people can compete effectively with others within the local tourism industry. Leading on from the foregoing, Kenya's sex trade is no one-way ticket out of poverty, but earlier researchers on the subject have pointed out how important the business is for some (poor) members of tourist host communities, especially women, looking for economic advancement (Finnegan 1979; Cohen 1982, 1993; Cabezas 1999; Kibicho 2004a, 2004b, 2005b). It is sometimes combined with other survival strategies, more and more micro-enterprises (see Chapters 5, 6, 8 and 9).

Conclusion

Various factors have been identified as leading Kenyans into sex trade (see also Figure 3.4). The key reasons are unemployment, family problems, pleasure and adventure. Thus, knowledge of the reasons for being in the trade is useful for understanding the range of activities and the multi-faceted nature of CSW–customer relationships (Figure 1.2). It should therefore be noted that the regular police crackdowns on CSWs anywhere in Kenya are only ever temporary remedial action, and address symptoms rather than tackling the predisposing factors. Instead of mounting such crackdowns, local/regional governments should develop programmes to intensify local people's participation in the tourism industry by improving the position of the existing and would-be small-scale tourism entrepreneurs.

This chapter has discussed pertinent issues regarding CSWs in Kenya's coastal region. For instance, it looked at sex trade as a response to unmet needs, financial hardship and general poverty. Thus, socio-economic contexts should be considered thoroughly to arrive at a proper understanding of the trade and its manifestations. A a consequence, the chapter normalised at least some components of sex trade, thereby presenting a new way of looking at it. However, the chapter did not answer one pertinent question: What challenges do CSWs in Kenya's coastal region face as they conduct their business? This is the foundation of Chapter 7, 'Sex Trade in Kenya's Coastal Region: A Sword that Cuts with Both Edges'.

Chapter 7
Sex Trade in Kenya's Coastal Region: A Sword that Cuts with Both Edges

Introduction

Sex trade involves a strong element of luck (Graburn 1983; Kelly 1988; Cohen 1993; Sindiga 1999). It is like a sword that cuts with both edges, in the sense that it offers both opportunities for success and riches, and risks to life and limb (Cohen 1993). Chapter 6 examined how sex trade offers some CSWs opportunities to deal with socio-cultural and economic challenges. However, this examination remains incomplete without an evaluation of the risks the sex workers face as they transact their sex business. This chapter continues the discussion in the previous chapter. It deals with the second part of the questionnaire described in Chapter 6. Thus, it explores the major challenges CSWs in Kenya's coastal region face as they transact their sex business. By doing so, it serves as the link between tourism-oriented sex trade in Kenya's coastal region and Malindi Area (Chapter 8).

In Kenya, as in many other countries, sex workers are vulnerable to risks such as physical attack or even social stigma. Table 7.1 represents the CSWs' ranking of the risks they face while conducting their profession. It should be noted that the 'harassment by clients' and 'fighting over clients' options were introduced to the study by the respondents themselves under the item 'Others' in the research questionnaire.

Tourism and Sexually Transmitted Disease

Travellers have a long history of being vectors for the spread of diseases (Walkowitz 1980; Muroi and Sasaki 1997; Agrusa 2003). The caravans travelling from Asia to Europe via the silk route played a great role in the spread of the Black Death (bubonic plague) that killed over a quarter of Europe's population (about 25 million) in the fourteenth century (Agrusa 2003). Numerous other examples can be cited where travellers have spread new strains of flu around the world. A more recent example is Severe Acute Respiratory Syndrome (SARS), which was first diagnosed in China in 2005, then spread to a number of countries (notably Canada). Another recent example is the spread of influenza A (H1N1 virus), also known as swine flu in 2009. The first case was reported in Mexico, a popular tourist destination, then spread worldwide within six months. In general, tourists' movements from their home countries to new destinations have the capacity to

Table 7.1   Risks encountered by sex workers (*n* = 183)

Risk	Rank	No. of respondents	%
Sexually transmitted diseases	1	165	90
Police harassment	2	170	93
Harassment by clients	3	157	86
Social stigma	4	161	88
Harassment by club operators	5	121	66
Fighting over clients	6	174	95

promote the transmission of viruses, especially those associated with STDs such as HIV-AIDS (AVERT.ORG 2004).

Health risks were the most feared threat to sex business in Kenya's coastal region by 90 per cent of the total respondents (*n* = 183) (see Table 7.1). Ninety-eight per cent of the CSWs identified HIV-AIDS as their most feared STD. This is quite similar to Muroi and Sasaki's (1997) findings in Bangkok, where CSWs were reported to be aware of the threats posed by HIV-AIDS to both them and their clients. This may be attributed to the regular anti-AIDS campaigns by the Kenyan and Thai governments (see also Kibicho 2004b). However, CSWs in Kenya described other STDs like gonorrhoea, syphilis, genital herpes and chlamydia as a 'normal cold'. In other words, the CSWs were not worried about contracting these STDs as they are treatable.

Condom Usage by Sex Workers

The global AIDS pandemic has changed the face of sexual encounters and practices. Initially, CSWs were blamed for the transmission of the virus into the heterosexual population in the developed world (Cohen 1988b; Leheny 1995; Agrusa 2003; AVERT.ORG 2004). However, empirical studies in a number of countries have shown that CSWs are fastidious in the use of condoms when offering penetrative sexual services (Robinson 1989; Bauer and McKercher 2003; AVERT.ORG 2004; Kibicho 2004a, 2004b, 2005b). They are good peer educators on safe sex practice. Sex workers in Kenya's coastal region, for instance, are aware of the importance of condom[1] usage as a way to prevent HIV-AIDS infection (see Table 7.2).

The majority of CSWs in Kenya's coastal region felt that condom usage while offering penetrative sexual services prevented STD infection (mean score = 4.60; Standard Deviation = 0.9). As a result, 85 per cent of the sex workers 'Always' used condoms when offering commercial sexual services. Only 3 per cent of the total CSWs answered 'Never' to the question 'How often do you use condoms when offering penetrative sexual services?' However, 64 per cent of the CSWs

---

1   In local sex trade circles, condoms are referred to as 'CDs' or 'socks'.

Table 7.2 **Responses regarding condom usage and their role in reducing the spread of STDs (n = 183)**

Statement	Mean	SD[a]
What are the possibilities of condoms preventing you contracting STDs? (Scale: 1= Very low; 2= Low; 3= No idea; 4=High; 5=Very high)	4.60	0.9
How often do you use condom(s) when offering penetrative sexual services? (Scale: 1= Never; 2= Sometimes; 3= Always)	2.80	1.1

Note: [a] Standard Deviation.

(n = 183) had on several occasions had sex with their clients without condoms. In all cases, the clients requested to have sexual intercourse with the CSWs without condoms. The extra amount of money the customer was willing to pay for non-use of a condom during intercourse was the principal motivator in the CSW's decision. A female CSW in Malindi Area (Figure 6.1), for example, reported that she charges the following fees for sexual services without a condom: K£53 for a 'short time' (or 'one shot') instead of K£25, K£158 for three shots instead of K£75, and K£368 for over three shots instead of K£175 (see Table 9.3 for exchange rates).

From the interviews with female CSWs, it was evident that many men paid more for sexual intercourse without using condoms. These men gave varying reasons for their insistence on this. Joel (fictitious name), a local tourist in Mtwapa (Figure 6.1), offered the following:

> I cannot imagine myself paying for sex, then I use a condom during the intercourse. To me, it makes no sense .... It interrupts the flow of sex, I want to feel the flesh, you know *nyama kwa nyama* ['flesh to flesh']. And in fact carrying CDs [condoms] all over the place means you are promiscuous. I am not promiscuous, I am doing it [buying sexual services from a CSWS] only for a short time. If a CSW insists that I use one, I wait until when the lights are off, I remove the damn thing. She realises that I did not use it [the condom] when I am through with my business .... You have to be sharp to maximise your satisfaction [he concludes with a misplaced sense of self-pride].

After a critical evaluation of Joel's reasoning, one is tempted to conclude that some of the CSWs' clients are ignorant of the risks associated with non-usage of condoms during sexual intercourse.

This evidence has huge implications for the transmission of HIV-AIDS and other STDs by the CSWs' clients to their regular partners. Further, unprotected sexual intercourse facilitates the spread of STDs to and from countries/areas of origin and tourism destinations (Mulhall et al. 1994).

Aside from the appalling human costs, the spread of such diseases, especially HIV-AIDS, both to the customer and the CSW, has negative effects on sustainable economic development in general, and sustainable tourism development in particular (see also Chapter 9). The affected families and individuals expend enormous resources (financial, time and so on) taking care of the sick persons. Young Kenyans, for example, drop out of school to take care of their bedridden parents. With no employment opportunities available to them, the children concerned have to look for other means to generate money to pay for their parents' medication and purchase food for the rest of the family. Some of them end up trading in sex. This exposes them to the risk of contracting STDs. Some unlucky ones get infected with HIV-AIDS. This chain-reaction kind of spread of the AIDS virus devastates the entire family, their neighbours, and sometimes the whole village (Kenya 2002b). Based on this fact, therefore, tourism policymakers, tourism operators and tourists (where possible) will need to take a long, hard look at the impact of CSW–customer interaction in the informal sector, particularly the effects of the reported high levels of spread of HIV-AIDS in Kenya's coastal region and in the country at large.

Apart from STDs, some CSWs also suffer from mental health problems. This is illustrated by Jayne's (not her real name) desperation. She observed:

> I found myself dancing in a Malindi nightclub at the age of 11 years .... I have no idea how this happened. What I know is that I have had different kinds of clients [for sexual services], both foreigners and Kenyans. I tried to commit suicide but it did not work, so I turned to drugs. I want to die before my next birthday. I hate this job [sex trade]! I hate my body! I hate myself! [She bangs the table with her head then burst into tears.]

Sex Trade-related Health Problems: Possible Solutions

Since CSWs have large numbers of sexual partners, sex trade has often been associated with the spread of STDs (see also Chapter 9). Typical responses to these health problems include (1) banning sex trade completely, and (2) introducing a registration system for CSWs (see also Chapter 2).

According to the first school of thought, any form of trading in sex or any activity which supports such a trade, either directly or indirectly, should never be tolerated. Tough legislative measures should be put in place or enforced (where applicable) 'to protect the larger society from this moral decay' (Mwangi 1995: 4). This group thinks that registering CSWs is condoning a trade which should not be countenanced by any self-respecting society (see also Chapter 8) (Mwangi 1995; Sindiga 1996; Kithaka 2004). The second school of thought feels that the first measure is counterproductive. This feeling is based on the fact that banning sex trade tends to drive it underground, thus making treatment and monitoring more difficult (see also Chapter 2) (Stanley 1990; Thomson and Harred 1992; Ryan and Hall 2001; Jago 2003). As a result, all CSWs should be recognised through a registration system which mandates health checks and other public health measures. Such measures should include educating the CSWs and their clients to encourage the adoption of preventive measures such as use of condoms. Thus, this stand can be viewed as a harm-reduction policy.

The encouragement of safe sex practices, combined with regular medical checkups, has been found to play a significant role in the reduction of the spread of STDs (Chapman 2005). In consequence, sex trade ceases to be a major vector of such diseases when safe sex practices are applied consistently (Agrusa 2003). It is a known fact that in regions where sexual precautions are not practised for religious, cultural, political and/or economic reasons, as in most African countries, sex trade appears to be one of the principal means of transmission of STDs, including the life-threatening HIV-AIDS (Ndune 1996; Mungai 1998). As a result, the prevalence of HIV-AIDS in Kenya, especially among CSWs, has been the subject of significant media and academic attention. In addition, Kenya hosted the 13th International AIDS Conference in 2003. Nevertheless, this does not seem to have dissuaded a significant number of tourists from partaking in sex with local CSWs.

Today, awareness on safe sex practices is high among most CSWs (Kibicho 2004a, 2005b). After the enactment of the Kenyan government's first five-year plan to combat the spread of the HIV-AIDS epidemic in 2004, for instance, condom usage during commercial sexual relations increased enormously, from 30 to 90 per cent (see Chapter 9) (Kenya 2006).

It begs reiterating that presently, HIV-AIDS has no cure or vaccine. Among other means, HIV-AIDS is transmitted by transfer of bodily fluids through sexual intercourses and intravenous drug use. Sexual intercourse, both heterosexual and homosexual, is the main mode of HIV transmission in the world today (AVERT. ORG 2004). This means transfer of HIV is therefore through conscious behaviours, if those engaging in sexual relationships are capable of making responsible or mature decisions. With the absence of a vaccination or an effective drug, the sole way to reduce the spread of HIV infection is through educating people (especially those who are sexually active) on how to employ effective preventive measures such as condom use (see also Chapter 8). In Kenya however, contentious debates arise about whether sex education and the availability of condoms will increase sexual activity among the youth and result in even higher rates of STDs (Kibicho 2007). But the dramatic differences seen in the spread of HIV-AIDS, pregnancy and abortion when we compare Kenya and other developed countries, where such programmes are funded and executed by governments, does not support such beliefs.

The Netherlands and Germany are examples of countries which have succeeded in controlling the spread of sex trade-related STDs (Agrusa 2003). Both countries have emphasised the hygienic aspect in their legislation by rigidly enforcing periodic medical examination of CSWs (see also Chapter 2). Sex laws in these countries provide free but compulsory hospitalisation for CSWs who are found to be infected with STDs (Hobson 1990; Agrusa 2003; Chapman 2005). This emphasis on regulation rather than suppression has resulted in a marked decline in the spread of STDs (Corbin 1990; Hobson 1990). It has also minimised the level of corruption, as fewer CSWs have to bribe law enforcement officers for 'favours' (Kibicho 2005b).

HIV-AIDS and Kenya's Tourism Development

In Kenya, the dynamics of the HIV-AIDS epidemic varies according to region (Kenya 2006).[2] The first cases were diagnosed retrospectively in Nairobi City/ Province in 1986 (Kenya 1987, 1989, 2006). As in many developing countries, heterosexual transmission predominates. Since 1990, the epidemic among drug addicts has particularly affected the coastal region (PTLC 1998). Throughout Kenya, the proportion of women among HIV-AIDS cases has increased over time, and now represents about 55 per cent of all cases (Kenya 2006).

Realising the dangers posed by HIV-AIDS towards national development, Kenya's government has mobilised enormous resources to fight against the spread of STDs in the country (Kenya 2002a, 2004, 2006; Kibicho 2004b). The government has implemented a number of techniques to curb the spread of HIV-AIDS, which is said to be killing a minimum of 700 Kenyans daily (Kenya 2006). Such techniques include operating free HIV-AIDS testing programmes, and distributing condoms at almost all public-meeting places, such as bars, restaurants, airports and hospitals. A committee on AIDS has been put in place under the chairmanship of the president of the Republic (Kibicho 2004b; Kenya 2006). The sole aim of this commission is to combat the spread of HIV-AIDS.

This fight is currently being carried out on two fronts. First, the Health Ministry co-ordinates HIV-AIDS screening programmes and provides anti-retroviral drugs to patients. Second, the Internal Security Ministry, notoriously, organises regular police crackdowns on CSWs in all main urban centres.

National AIDS awareness campaigns The Ministry of Health is responsible for disseminating general information on health. Its education programme has three objectives:

1. to ensure that the general public is regularly and properly informed;
2. to organise awareness campaigns for the general public and for targeted populations;
3. to co-ordinate prevention and health-promoting interventions in the field, led by various public and private partnerships.

In September 2002, the first national prevention campaign appeared on all television channels in the form of an advertisement: '*Pamoja tuagamize ukimwi*

---

2 Administratively, Kenya is divided into eight provinces which are then divided into districts. Districts are further divided into divisions, which are then divided into locations.

– Let's unite against AIDS'. During 2002–2003, apart from six campaigns centred on the disease (patient testimonies, solidarity, encouraging HIV testing and access to health care), an additional four ministerial campaigns promoting condom use were implemented. The general strategy was based on a distinction between HIV-AIDS prevention messages and information about condoms. The campaigns promoted condom use by trying to improve its image, by popularising them and making information on their use widely available. The HIV-AIDS message aimed to prevent any stigma being attached to populations 'at risk', encouraging solidarity and fighting against the attempt to deny that a problem existed, without generating panic in the population (Kenya 2004a). The messages were underpinned by a morality that stressed individual and collective responsibility.

The priority target populations are chosen according to how representative within the epidemic they are, or because of their particular vulnerability, or due to their importance in a prevention campaign. These populations include seropositive persons, homosexuals, drug users, multi-partner heterosexuals, adolescents, prisoners and CSWs (Kenya 2006). Despite encouragement from the Ministry of Health to promote prevention programmes for these populations, action is mainly directed at adolescents and drug users attending therapeutic clinics.

However, the reality on the ground is full of contradictions. For example, only a minority of these programmes have been directed at other 'priority' populations like homosexuals and the CSWs (Kibicho 2004b). Further, the government has opposed the setting up of a needle exchange programme despite a sharp increase in the number of intravenous drug users. A needle exchange programme is a project through which the government would supply the intravenous drug addicts with clean/new needles. A similar programme which had been established by a private drug rehabilitation centre in Malindi in 2003 failed due to lack of government support (Kibicho 2004b). The government fears that such a programme might encourage more drug users to start injecting themselves (Kenya 2006). However, this is not what these programmes do, and the government ought to know that. Safe injection sites, for example, can be places for addicts to learn about detox programmes, places for social workers to get the message of well-meaning policymakers out to the people who need to hear it – people who spend most of their time in alleys and stairwells and do not watch a lot of public service announcements. It is fair to say that a needle exchange programme does not send a message that is okay to (ab)use drugs. It reduces the chances that intravenous drug users will spread HIV-AIDS and hepatitis to the general community. It begs noting that most of these addicts share needles, as they cannot afford to buy clean ones. This creates a new dimension in the fight against HIV-AIDS in the country, and in Africa in general.

The long-term health implications of the spread of HIV-AIDS and other STDs are enormous, and represent a potential time bomb for Kenya's economic development. Unless the government takes bold, firm and decisive action aimed at overcoming deep-rooted cultural attitudes towards sex, not only will the broader tourism industry be damaged, but the human base of Kenya's development will be undermined.

Harassment of Sex Workers by Law Enforcement Agents

Sex workers' harassment by police officers was placed second by 93 per cent of the respondents (Table 7.1). It should be noted that the more visible the sex trade, the more vulnerable CSWs are to harassment and exploitation by police officers. This again explains why the majority of CSWs prefer the soft-sell technique when trading their sexual services (see Chapters 5 and 9). The most prevalent form of harassment is demand for payments by some corrupt police officers. Sometimes, security officers intimidate female CSWs in order to get free sexual services. Other forms of mundane abuse against CSWs carried out by Kenya's security personnel include raids, (gang) rapes, beatings, extortion and confiscation of property (see Kibicho 2004a, 2004b, 2005b). Such treatment is probably what led Naibavu and Schutz (1974: 65) to take a rather mercenary role in their appraisal of sex tourism in Fuji's national economy. They observe that sex trade:

> meets the criteria laid down in the Government's Development Plan more fully than almost any other industry. It is fully localised industry which gives employment to unskilled CSWs for most of whom no other jobs are available [see, for example, Figure 9.2]. It requires no investment of capital, yet it brings in large amounts of foreign exchange with a minimum of leakage back overseas. Therefore sex workers should be respected and be allowed to transact their businesses without intimidation by the law enforcers ....

Twenty-five years later, while commenting on the situation in the Dominican Republic, Cabezas (1999: 105) was of a similar view. He noted:

> The circulation of CSWs in the economy brings profit to many, from the transnational hotels and airlines to the small street vendors who sell hair ornaments ... hotel managers, taxi drivers ... and many other intermediaries traffic CSWs and usually procure a cut of their earnings. The police, the state, and the local and transnational enterprises are all aware that sex has a market value ... even while they are proclaiming that sex trade is immoral.

Another area where growing recognition of the plight of CSWs in Kenya may be affecting change is in the area of sex worker's exploitation through the contradictory legal system. A major reason for the routine exploitation of CSWs is the contradictions that exist between legislation and enforcement of sex trade-related laws (see also Chapter 2). In fact, some law enforcers are aware of the enormity of the problems inherent in this legal situation. For example, Roy (not his real name), a police constable in Malindi, felt:

> Something needs to be done to ease this legal confusion. Unfortunately, I am only a small 'fish' in the police force. I sympathise with these ladies. When it comes to protecting their basic (human) rights, I feel very much like a one-

legged man in an ass-kicking contest. I am from a very poor background, I do not rejoice at this. Trading in sex is the only 'crime' in Kenya where two people/adults do a thing mutually agreed upon and yet only one, the female partner, is arrested and punished. It is kind of an odd feeling to see poor people in such difficult straits.

This contradiction renders sex workers vulnerable to exploitation by their clients and law enforcers.

Related to the foregoing, however, it is worth underlining the fact that legislation or decriminalisation of sex trade should not be considered as a panacea or immediate answer to all its associated problems. However, it can be a means of securing better support mechanisms in terms of health, protection and self-(re)assessment by CSWs.

On some occasions, however, CSWs develop symbiotic working relations with local police officers. Sex workers pay a 'commission' to the officers on duty in order to be allowed to operate in a given area at a given time. This said, it begs noting that in Kenya, and in the coastal region in particular, policing of sex trade is reactionary, led by local residents' complaints. It thus varies over time and place, with police sometimes launching zero tolerance-style crackdowns on sex trade in areas with persistent complaints. During this kind of operation, a higher than usual police presence is put in place for a pre-defined period (Sindiga 1995; Ndune 1996; Kibicho 2005b). However, some police units allow 'unofficial' sex trade tolerance zones during the crackdown period. Sex workers are only allowed to work in these zones and at particular times. None the less, this arrangement has three major shortcomings. First, CSWs operating in these zones pay an illegal daily fee to the security officers, which reduces their net earnings. Second, because this is unofficial arrangement, it is difficult to communicate to all CSWs and other police officers the locations declared as 'tolerance zones'. Third, changes in senior police staff and/or renewed complaints by local residents lead to sex trade operation zones being changed or closed with no notice. These factors put the CSWs in a vulnerable position, as they cannot plan for their future since they work in unreliable and highly unpredictable conditions.

Mistreatment of Sex Workers by their Clients

Despite the fact that this researcher had overlooked this option and thus omitted it as an item in the questionnaire, 86 per cent of the sex workers ranked it as the third highest risk (Table 7.1). Such harassment entails both physical and non-physical attacks, mainly through non-payment for sexual services already delivered.

In terms of physical attack, Eva (fictitious name), a CSW from Mtwapa (Figure 6.1), described an ordeal she experienced when one of her clients, with the help of his two colleagues, literally inserted a Coca Cola bottle into her private parts. Eva completed the story:

> They disappeared when I became unconscious, of course, without paying for my services. I am still suffering from that barbaric act, both emotionally and physically .... Today, I walk with difficulties. Nevertheless, here I am ... still surviving by the Grace of God and still serving the same unappreciative clients. What can I do to him ... to the clients who did this to me? My hatred toward them is only hurting me, so I have no choice, I have to forgive and forget. My other alternative is death [she adds amid sobs].

Ironically, she could not report the matter to the police since she is involved in an illegal profession (see Chapters 2 and 6). This is in conformity with Cohen's observation that 'these women sometimes pay with their lives' (Cohen 1993: 160).

As far as non-physical harassment is concerned, Liz (pseudonym), another female CSW in Kilifi (Figure 6.1), recounted how she once spent a night with a client who, in her words, terrified her throughout the seven hours they were together. She explained:

> He had a long, shiny and sharp knife .... He kept it under the mattress of the bed we were sharing. He constantly reminded me that he is a 'professional' user of the tool and that he was ready and willing to use it on me if I disobeyed his orders. He demanded sexual services which I have never imagined – awful things! I feel ashamed whenever I think of what he forced me to do that night. It was extremely humiliating – I do not want to talk about this [she covers her face with her palms]. Anyway, at the end of it all, he did not pay for my services [commercial sexual services]. There are many clients like that, but instead of risking my life, I prefer leaving them alone. Good Lord will punish them for me ... [she laughs].

Related to the foregoing, Jacques (not his real name), a 39-year-old French male tourist in Shimoni (Figure 6.1), gave an accurate description of how non-violent harassment of CSWs is carried out:

> I have been screwing them with no pay .... Sometimes I pretend that I do not have money, and the girls [CSWs] can do nothing to me. They cannot report me to the police .... At times I give them a dollar [US$1] after a quick fuck on the beach ... but now things are becoming difficult for me. They [CSWs] no longer want to serve [offer sexual services to] me. Occasionally, a few of them accept to sleep [have sex] with me, but I have to pay for the services upfront. This weakens my bargaining power as I am paying for untested service – a dream! The majority of them avoid me. I think they have come to know my strategy. In fact, I am planning to shift from here. I might be heading to Mombasa, where the local gals [CSWs] do not know me.

Social Stigma Attached to Sex Workers' Survival

Social stigmatisation was ranked fourth by 88 per cent of the sex workers ($n = 183$) (Table 7.1). Fifty-four per cent of these sex workers were of coastal origin. The majority of the CSWs from upcountry had no relatives in the region, thus they appeared to be indifferent to social pressures, unlike their coastal counterparts, who were known by relatively many people within their areas of operation. All those who mentioned this as a problem preferred to operate in locations far from their homes or where the local residents did not know them.

In response to the general stigmatisation of sex trade, and sex workers in particular, Sophie (fictitious name), a CSW in Malindi, asked:

> Do those against us [CSWs] bring money home to feed us? Would they take care of our dependants? I am sure they would do the same if they were in our positions. Do they think we are in the sex trade for fun? ... because we love having sex with strangers, some of whom do not even brush their teeth ...? They are mistaken. We are working to earn a living just like any other hardworking Kenyans. Yeah, the government continues with its empty political rhetoric on poverty eradication. You know, 'power-point presentations' filled with personal opinions and based on speculations. They [government officials] can continue with their baseless 'preaching', but I can assure you that we cannot combat poverty unless we first fight the social injustices of the policies we have collectively adopted. We [CSWs] do not want charity. We want opportunities. We have created our own opportunities in the sex industry. The government has got to take off its ill-founded ideological blinders. It is high time the local governments, and most importantly Kenyans, start looking at us more positively .... We deserve this recognition. We deserve respect..

Not only did Sophie speak good English, she also spoke German and rudimentary Italian. She was learning French so that she could have an edge on the 'French market', as there had been an influx of French male tourists seeking African women in Kenya's coastal region, Sophie reported. When asked what she did before entering the sex trade, she smiled, ignored the question, and mumbled that all this talk was ruining her business. Sophie said she was working that day for a K£25 phone card to call her sick mother in Central Kenya.

Sophie's comments were echoed by other CSWs throughout Kenya – women work in the sex industry because they are responsible for taking care of their children and their extended families. In times of shortages of food and other necessities, it is women who have to provide them. 'We therefore have to carefully strategise so that we can come out of these stressful situations,' Sophie noted. In any case, in Kenya's patriarchal family system within a patriarchal society, it is women who must either make do with less, or find ways to earn more. In the struggle to survive as well as to keep their families alive, many Kenyan women will turn to whatever means available. As men leave their families and their localities in search of work and better lives in more promising areas, 'women are left with the burden of providing for themselves and for their families in a society that pushes them into a way of life that it shuns. It is contradiction *par excellence*,' Sophie concluded (see also Chapter 9).

In other words, women bear the burden of managing poverty on a day-to-day basis. Whether they live alone or with a partner, on social benefits or low earnings, it is usually women who are responsible for making ends meet, and for managing the debts which arise when they don't. Indeed, the lower the household income, the more likely it is that this responsibility will rest with women (McLeod 1982). As unemployment levels rise and as social benefits are 'cut', women are increasingly caught in a daily struggle to feed and clothe their families – usually at considerable personal sacrifice (Cohen 1982; Burton 1995; Brown 2000).

Morally and ideologically, the responses to sex trade and the sex industry in general are historically rooted in a double standard of morality (Walkowitz 1980; McLeod 1982; McLintock 1992). This double standard divides women into two – women who are expected to be loyal to their husbands and to be good mothers, and others (see, for example, McLeod 1982; McLintock 1992; Muroi and Sasaki 1997). Some Kenyan women accept this double standard, either willingly or unwillingly (Mungai 1998). Some tend to believe that having extra-marital affairs, for instance, is natural for men, thus wise women do not complain about it. Those who are in the category of 'good wives' find it easy to accept the norm that women who sell their bodies are a different type of woman, in a category of 'whores'. This is further reflected in the laws relating to soliciting and kerb-crawling in Kenya (Jackson 1986). In fact, it is embedded in the whole socialisation process (see Caplan 1984). As a result, any woman is vulnerable to the 'whore' stigma because of life experience, sexist abuse or ill fortune (Ackermann and Filter 1994; Cabezas 1999). It is worth noting that female CSWs' vulnerability to stigmatisation rests on their relatively poor power position. At the same time, when CSWs are effectively stigmatised, that reinforces their overall subordination and makes it more difficult for them to achieve desired goals. Thus, stigmatisation becomes self-perpetuating or snowballing in its impact (McLeod 1982; Mensah 2005).

In relation to the foregoing, Nancy (not her real name), a female CSW in Mombasa, gave a double standard stigmatisation story:

> I have lost many friends. They look at me (very) differently ... it bothers me ....
> Sometimes I think, 'Fucking hell, I am an outcast, a slut, a social misfit' .... Yes, I am a lot of things, but I am not an outcast. I have two faces ... two different lives – work life and private life. One evening, my three friends sat in my house watching evening television news. That evening, the news presenter belaboured on the measures the government was taking to 'ruthlessly deal' with the tourism-related sex trade, especially in Mombasa and Malindi. One of them commented while pointing at the images of CSWs on the TV screen, 'Look at them dirty prostitutes ...'. I spontaneously responded to his comments by reminding him that I am a sex worker and this is my house, my television ..., all paid for through

trading in sex. Everybody looked at me with a remarkable bewilderment ....
Everyone stopped talking to me. A few minutes later, they all left in unexplained
hurry. I was so terrified the whole evening .... I had horrible nightmares ....
In the morning, I just counted my previous night's earnings – it was my only
consolation. We are living in a too judgmental and a very unaccommodating
society. It is kind of name-calling at its very basic level!

However, as noted earlier, the bottom line is that most CSWs accept sex trade as a means of survival – survival for themselves and their families back home (see Chapters 6 and 9). Even though they cannot escape feelings of shame, they have pride in the fact that they contribute to the sustenance of their families.

Other Sex Trade-related Risks

Other risks mentioned by respondents include harassment by club operators (66 per cent of the CSWs) and fighting over customers among the sex workers (95 per cent). The former was ranked fifth, while the latter was ranked last by the respondents (see Table 7.1). Many cases of fighting among CSWs involve Kenyan women and those from other Eastern Africa countries such as Burundi, the Democratic Republic of Congo, Rwanda, Somalia, Tanzania and Uganda (in alphabetical order). More often than not, such fights end up being inter-group conflicts, as members defend their 'sisters' (see also Chapter 9). Sonia (not her real name), a Kenyan female CSW who has been working in Mtwapa (Figure 6.1) for the better part of the last decade, justified these fights by saying:

These Rwandese, Ugandans ... are stealing our business. They do not respect
prices ... they are taking clients for much less money. It is unfair competition.
Does it mean that there are no men who can purchase their sexual services back
in their countries? They should go home and look for men there.

Such situations are aggravated by the fact that, as in most informal economic sectors, supply exceeds demand in the sex trade profession, as there are more sex workers than potential customers. Further, sex trade in Kenya's coastal region is subject to booms and slumps in the tourism industry. For instance, 64 per cent of the CSWs in the region felt that late 1997 was the worst period for their profession. This was when tourism performance in the region was at its lowest ebb, with bed occupancy for many tourist hotels nose-diving from 70 per cent to as low as 20 per cent; some hotels closed down (see Chapter 6 and Table 6.1). The situation increased the number of CSWs and at the same time reduced the number of customers available for their sexual services, as international tourists cancelled their trips to the region (see Chapter 9). Suzie (not her real name), a CSW in Malindi, explained this situation in the following way:

When there are few tourists coming to Malindi, the size of the market for our services shrinks. And since the number of sisters [CSWs] remains constant, and in some cases increases, due to new entrants into the trade, the supply outstrips the demand. As a result, there is normally a stiff competition. Only the smart ones survive. It is survival of the fittest at its best. On the contrary, when the tourism industry is performing well, we all get our shares of the tourist dollars with less competition and struggle.

This confirms some level of linkage between sex trade and the tourism industry's performance in Kenya's coastal region.

A much underestimated and under-studied social problem for CSWs is peer abuse. Peer abuse, like peer interaction, is socio-culturally situated in CSWs' lives. It varies over time and space. In other words, it is produced and resisted in various ways across social, cultural and gender planes. Consequently, a socio-cultural approach to peer abuse draws our attention to the fact that female CSWs face a rougher 'world' than their male counterparts, reporting greater exposure to gangs, drugs and rape, and more fear of violence from their clients (see, for example, Eva and Liz's stories discussed earlier in this chapter) (see also Hanmer et al. 1989).

Conclusion

Based on the arguments put forth in this chapter, there is no doubt that if Kenya's government is serious about dealing with the growth of sex trade in general and arresting the spread of HIV-AIDS in the coastal region in particular, major emphasis has to be placed on supporting mechanisms for those who are forced to use prostitution as a means of employment. From a sustainable tourism point of view, the government should initiate new forms of tourism to absorb sex workers willing to quit the sex trade in favour of an alternative source of income (see Chapters 6 and 9). Such projects should consist of small-scale, dispersed and low-density tourism developments organised by tourist host communities, where it is hoped they will foster more meaningful interaction between tourists and local residents. By stressing small-scale, locally owned operations, it is anticipated that tourism will increase the multiplier and spread effects within the host community and avoid problems of excessive foreign exchange leakages (see Chapters 3 and 9).

However, this chapter has not examined the specific roles played by MCSWs in sex trade. This examination is justified by the fact that Kenya has gained unprecedented popularity as a romance tourism destination – a favourite for female sex tourists seeking sex and romance (see Chapter 5) (Kibicho 2004b). Therefore, Chapter 8 – 'Sex Trade in Malindi: Roles Male Sex Workers Play' – takes up the challenge of analysing activities by MCSWs in an undeniably homophobic tourism destination area.

Chapter 8
Sex Trade in Malindi: Roles Male Sex Workers Play

Introduction

This chapter deals with a subject which has historically received little attention in tourism studies in Kenya, and in Africa in general – the relationship between tourism and the development of the male sex trade (Kibicho 2004b). Male sex workers are often excluded from social studies and interventions which target CSWs (see also Chapter 6). This fact is due, at least in part, to the social marginalisation of (male) homosexuality (Mungai 1998). Male sex trade, like occasional female prostitution, is sometimes however difficult to distinguish from simple flirting between homosexuals. In other words, one must distinguish between 'professionals' making a living only from their sex work, and boys (often young) who work occasionally to buy a Walkman, a leather jacket or other trinkets of Western values.

This chapter thus looks at these links within the reality of Malindi Area, where rural overpopulation and the demands of an emerging and diversifying urban economy are rapidly transforming economic and social relationships. The main aim of this chapter is to offer something more than simply the descriptive anecdotal evidence provided by past researchers. Therefore, it seeks to provide structures of analysis to help and locate tourism within the wider context of the sex trade in Malindi. Such an understanding will certainly provide some valuable quantitative insights into the data-deficient aspect of the relationship between tourism and male sex trade in Kenya, and could thus be of use to further study.

By examining the importance of tourism in Malindi's local economy then investigating the relationship between this industry and sex trade, the first part of this chapter lays the foundation for an objective analysis of male sex trade as a form of entrepreneurship. This chapter therefore advances the exploration of the links between sex (trade) and tourism in Kenya in the proceeding chapters. It also sets the scene for the discussion on the influence of the tourism industry on the operations of CSWs in Malindi in Chapter 9.

Tourism Development in Malindi Area

Locational Context

The town of Malindi[1] serves as the headquarters of Malindi District.[2] It is located between latitudes 3° 15' and 4° 00' south and longitudes 40° 50' and 41° 43' east, which is 125 kilometres north of Mombasa City (Figure 6.1). With a population of about 345,000 inhabitants, Malindi is the second largest of Kenya's coastal towns, after Mombasa (Kenya 2002b). The area has a coastline 52 kilometres long (Kenya 2002a, 2002b). It is generally hot and humid throughout the year, with a mean daily temperature ranging of 21–30°C (PTLC 1998). Mild temperatures characterise the peak tourism season, which runs from September to March. In addition, warm water temperatures throughout the year allow tourists to participate in various water-related sports – swimming, wind-surfing, sport-fishing and water-skiing – during both peak and off-peak seasons (see Chapters 3 and 6). The importance of tourism to Malindi's economy can not be overstated. It is the largest component of the local economy, contributing over 85 per cent of the area's economic activities (Peake 1989; Kenya 1993, 2002b; PTLC 1998; Kibicho 2003).

Size of the Economy: Urban versus Rural Areas in Malindi District

Migrant workers from rural areas usually enter an urban area of their own country before moving to richer countries (Archavanitkul and Guest 1994). In normal circumstances, people move to places where they hope to get employment opportunities (see, for example, Phongpaichit 1981; Findley and Williams 1991; Muroi and Sasaki 1997). The huge difference in economic wealth between urban and rural areas is often cited as the principal incentive for rural–urban migration. This theory perfectly fits the situation in Malindi District. In 2002, for instance, the economic structures of the three divisions of Magarini, Malindi Rural and Marafa differed considerably from that of Malindi Town (see Table 8.1) (Kenya 2002b, 2006).

According to Table 8.1, the agricultural sector accounts for 43–50 per cent of the total GDP in the three divisions, while in Malindi Town the figure is only 4 per cent, compared with the 84 per cent shares of services, mainly in the tourism industry. There is a considerable gap between the economic wealth of Malindi Town and that of the remaining divisions (Kenya 2002b). Paradoxically, while only 14 per cent of the total population of Malindi District resides within the Malindi Municipality, it accounts for over half of the total GDP (Kenya 2002a, 2002b, 2003). It is evident from Table 8.1 that Malindi Municipality has grown

---

1 In this book, the terms Malindi Municipality, Malindi Town and Malindi Area are used interchangeably.

2 Malindi District comprises four divisions: Magarini, Marafa, Malindi Rural and Malindi Town.

Table 8.1 Gross Domestic Product by region – Malindi District[a]

	Magarini	Marafa	Malindi Rural	Malindi Town[b]	Totals
Agriculture	63,220	55,300	60,241	21,071	199,832
Industry	20,012	18,690	19,006	49,850	107,558
Tourism-related services	56,830	54,481	40,610	382,806	534,727
Total GDP	140,062	128,471	119,857	453,727	842,117
GDP (%)	16.6	15.3	14.2	53.9	100
Population (%)	36	21	19	14	100

Notes:
[a] Prices in K£ million (US$1 = K£3.8) are as of 2002.
[b] Malindi Town refers to the area designated as Malindi Municipality.
Source: Kenya (2002b: 19).

in economic terms, leaving the neighbouring divisions behind. This induces local people to migrate from rural areas to work in the urban service sector, mainly in the tourism industry, with the intention of finding a better standard of living. In other words, these migrations are due to the economic imbalance between rural/peripheral and urban/core areas (see, for example, Findley and Williams 1991).

Many earlier researchers have used the core–periphery theory in analysing tourism impacts in destinations in the developing world (see, for example, Turner and Ash 1975; Britton 1982; Brown and Hall 2000; Scott 2000). Most of these studies emphasise the various forms of dependencies of the periphery, the dominated and marginalised areas, on the metropolis. The peripheries are characterised by a comparative lack of innovation. As a result, new products, new technologies and new ideas tend to be imported rather than developed within the periphery (Brown and Hall 2000; Scott 2000). While elaborating on the core–periphery theory, Britton (1982) identifies four principal mechanisms critical to the asymmetric interactions between the dominating 'centre' – for example, Malindi Area – and the dominated 'periphery' – for instance, the three divisions in Malindi District. These mechanisms are exploitation, penetration, fragmentation and marginalisation. This does not however, explain why many new immigrants from the periphery areas chose sex trade as their profession on arrival in Malindi. The examination of monthly pay by different occupations in late 2005 in Table 8.2 reveals at least one of the reasons. Compared with other jobs, local people who work as CSWs or masseuses have far higher earnings. This is a major factor inducing them to move from their local areas to work in the sex industry in Malindi Town. In any case, they would not come to Malindi Area if there were no pull factors.

Table 8.2   Monthly pay by occupation in Malindi Town, 2005

Occupation	Monthly pay (K£)[a]
Housemaid	50
Waitress	150
Construction labourer	375
Factory worker	450
Beauty salon	500
Clerical worker	750
Secretary	1,000
Masseuse	2,250
CSW[b]	3,000

Notes: [a] US$ 1 = K£3.8.
[b] Sex workers' income range: K£2,750 to K£7,500 per month (see also Chapter 9 and Table 9.3).

On a different note, poverty motivates people to migrate to cities from the confines of their rural villages in the hope of securing a better future for themselves and their families. Often, the only jobs these new immigrants can find are as receptionists, bartenders, waitresses, dancers, self-styled tour guides and street merchants. Owing to their lack of specialised/professional training and faced with severely restricted occupational opportunities, they find themselves on the street without work, food and shelter (see Chapters 6 and 7). For this reason, sex trade becomes a viable alternative for their survival. Agrusa (2003: 168) explains this scenario in the following words:

> For many people in the developing world, sex trade is one of the few realistic options for earning a decent income, particularly for young, uneducated women from rural areas. Given that the wages of factory and domestic servant jobs, the other two options open to women in this group, are so low[3] it is little wonder that they opt for sex trade.

It is discernible from the foregoing that men and women in Malindi, and in fact in other tourism destination areas in Kenya, engage in sex for cash in order to obtain money, from motives ranging from the desire to help their families to survive in conditions of extreme poverty to gaining additional earnings to spend on commercial products (see Chapters 2, 3, 5, 6, 7 and 9). Subsequently, sex tourism in Malindi Area can only be understood within a framework which addresses the varieties of structural inequality that occur within Kenya's coastal region and in

---

3   See Table 8.2.

the country at large. In any case, like tourism, sex trade in any tourism destination has a substantial local economic and social impact. Such a social change in Malindi seems to be interlocked with the increasing entry of locals from the rural areas into the urban centre, and their resulting vulnerable situation due to lack of employment opportunities (see Chapter 6).

As a consequence, the Malindi District development strategies currently being undertaken need to be re-examined, not only in the light of the enrichment of the different peripheral areas (divisions), but also with the intention of reducing the economic disparity within the district (Kenya 2002b). A micro-economic approach is more relevant in alleviating the hardship that women (and some men) are facing than a macro-economic perspective. As a consequence, development funds should be directed towards empowering individuals (see Chapter 6).

Birth of the Tourism Industry in Malindi

Development of modern tourism in Malindi dates back to 1931, when an 18-bed hotel was built as the base for deep-sea fishing off Malindi Bay (Peake 1989; Kibicho 2003, 2005a). This is where the world-famous reporter on sport fishing, Ernest Hemingway, stayed for several weeks in 1934. The hotel, classified as three-star, is still in existence under the name 'The Blue Marlin'. In the same year, the tourism industry was formally launched by the then District Commissioner, Sir Leo Lawfords. He later built a *makuti* (coconut palm frond) hotel in 1935 under the name 'Lawfords Hotel' (Kibicho 2003). The hotel still stands as a four-star establishment. From the local people's viewpoint, Sir Lawfords is considered a visionary for the type of planning he used in Malindi's development (PTLC 1998). It was based on a semi-master plan that included the establishment of three fundamental commitments: (1) strict control of land use, (2) development of recreational areas and (3) preservation of two square kilometres of marine reserve, currently known as Malindi Marine National Park (KWS 1997). His goal was to maintain the area's natural beauty by developing tourism in such a way that the destination's natural attributes would be enhanced rather than being destroyed (Peake 1989; KWS 1990, 1997; Kibicho 2003, 2006d).

However, with an ever-increasing (mass) tourism development (see Chapter 3), it was difficult to maintain the high quality set by Lawfords. By 1968, for instance, the resort had almost trebled its bed capacity to about 1,000 (PTLC 1998). On account of its pristine beaches and an influx of foreign investors, particularly from Italy, Malindi Area has experienced a tourism boom, especially in terms of hotel bed capacity. The area has a total of 5,700 beds, 42 star-rated restaurants, 28 tour operators and 517 beach operators (Kenya 2002b; Kibicho 2005b). It is estimated that there are 3,500 beds in private villas and homes used by tourists (Kenya 2006). At the same time, from 1995 to 2000, the total number of tourists to Malindi Area increased by almost 70 per cent, from 121,000 to 206,000 visitors (Kenya 2002b). Today, Malindi is a concentration of various types of tourist facilities, ranging from curio shops, cheap restaurants, bars, guesthouses and villas to luxurious hotels of international standard. As a result, the area has become a mass tourism destination, largely due to its attractions of sand, sun, sea and (to some extent) sex (Kibicho 2003, 2004a, 2004b, 2005a, 2005b, 2005c).

Research on Male Sex Workers

As noted elsewhere, a number of tourism researchers have investigated the relationships between male tourists and 'local women' in the visited destination (Cohen 1982, 1988a, 1993; Graburn 1983). However, little research has been carried out into the question of female clients for MCSWs, which is a relatively common feature in Third World tourism destinations (see Chapter 5) (Harrell-Bond 1978; Mathews 1978; Pruitt and LaFont 1995; ECPAT 2002; Afrol 2003; Kibicho 2004b). For instance, Trovato's (2004) study in Gambia reveals that while female sex trade is common, male sex trade among young Gambians is rampant with middle-aged Scandinavian women openly soliciting local young men. Similarly, there has been a significant growth in the number of Americans going to Jamaica to meet with the local 'beach boys' (Pruitt and LaFont 1995). Mathews (1978: 67) also reports that the: 'relationship between tourism and the sex trade in Barbados is not solely male-oriented. The industry also thrives on the alleged desire of white female tourists to have sex with black males ...'.

This development can possibly be attributed to the fact that the number of female tourists travelling unaccompanied by husbands or boyfriends is on the increase (see Chapter 5). Pruitt and LaFont (1995: 425) explain these changes by observing: 'travelling is a gendering activity as women tourists seek to expand their gender repertoires to incorporate practices traditionally reserved for men and thereby integrate the conventionally masculine with the feminine'.[4]

Although male sex trade is prevalent in Malindi, there is a substantial lack of systematic research on the trade. This is due to several factors, including:

1. the informal and illegal nature of the sex trade (see Chapter 2);
2. general unwillingness among police officers and government authorities to acknowledge its existence;
3. lack of academic interest in the subject.

As noted in Chapter 2, as in many other countries, it is illegal to be involved in the business of selling or buying sexual services in Kenya. Thus, male sex trade in the Malindi Area exists as a 'fuzzy' legal undertaking. This is because, while the trade is proscribed country-wide, it thrives in the area within a legal and social framework of constraint and limited tolerance. Moreover, laws regarding sex trade are in most cases poorly enforced by the authorities, not only in Malindi Area, but in the country at large (see Chapter 2) (Kibicho 2004b).

---

4   For more details, see Chapter 5.

The Relationship between Tourism and Male Sex Trade

Lack of official documentation of sex trade makes it difficult to generate an accurate sample frame or a complete list of the targeted population in Malindi (see also Chapters 6, 7 and 9). However, with the help of the Secretary of the MWA, it was possible to obtain an indication of the approximate number of sex workers in the Area (see Chapters 6 and 9). The MWA has perhaps the most reliable details on sex trade in Malindi, particularly in regard to its members, but less so with respect to non-members.

However, the sample frame employed in this study might have led to bias in the sample, as the population only consisted of the 184 MCSWs registered with the MWA. In addition, due to the general unwillingness of CSWs to share their profession's details with 'strangers' (non-CSWs), the author relied on their secretary's interventions when conducting a questionnaire survey as face-to-face interviews. But it is worth noting that such unwillingness could be professionally appropriate, as in some professions (like the sex trade), it requires considerable trust and a long-term relationship to discuss business-related issues (Alleyne and Boxill 2003). In any case, the very act of discussion can, in the more intolerant (government) systems, place individuals in danger. In addition, concepts of sex trade, sex workers' rights, sexual exploitation, (child) sex tourism, sex tourist, poverty and so on, seem not to be identified or accepted in the same way throughout the world (see 'Setting the Scene'). And finally, not everyone wants to talk about such 'emotive' issues with a stranger (Cohen 1982). Of course, there is nothing intrinsically wrong in that, and it would be disrespectful to compel sex workers to talk about issues which they do not care to discuss with strangers.

Using computer-generated random numbers, 160 MCSWs were selected to participate in this study. An equal number of questionnaires, 80 in total, were distributed during the peak and off-peak tourism seasons (see Chapters 3 and 6 for temporal distribution of these seasons) over a seven-month period, January–July 2002. Thirty-eight completed interviews were carried out during the peak tourism season, while the remaining 35 interviews were conducted during the off-peak season, resulting in a total sample size of 73. This resulted in response rates of 48 per cent for the peak season and 44 per cent for the off-peak season, which translates to an overall response rate of 46 per cent.

A structured questionnaire in Kiswahili and English was used as a primary tool for data collection. The questionnaire covered three main areas. The first part identified the demographic characteristics of the respondents, including age, marital status, origin, religion and level of education. The respondents' profiles are summarised below: 58 per cent of the respondents were in the age bracket 21–30 years; 70 per cent had a secondary level of education, while 72 per cent were from regions outside Malindi Area. Ninety-four per cent of the total respondents were unmarried, and 58 per cent subscribed to a Christian religion.

The second part of the questionnaire studied four issues:

1. the prices charged by MCSWs for their commercial sexual services;
2. the type of commercial sexual services the MCSWs offered to their clients;
3. the rate of condom use by respondents, using a three-point Likert-type scale, ranging from 1 (Never use) to 3 (Always use);
4. levels of dependency on tourism, using a four-point Likert-type scale, ranging from 1 (Highly not dependent) to 4 (Highly dependent).

The third part consisted of 14 closed-style items requiring the sex workers to rank their rate of agreement with particulars items. The 14 items sought the MCSWs' perceptions on the relationship between tourism and sex trade in Malindi Area. These items were measured on a five-point Likert-type scale, ranging from 1 (Highly disagree) to 5 (Highly agree) (see Table 8.3). Adoption of these items was based on three general observations by earlier commentators on sex trade. First, it has been reported that the tourism industry provides a ready market for commercial sexual services (Graburn 1983; Ndune 1996; Ryan and Kinder 1996; Brown 2000). Second, Naibavu and Schutz (1974), Hall (1996), Wilson (1997) and Phillip and Dan (1998) revealed positive impacts of sex trade on a tourism-driven economy; and finally, Cohen (1988a), Robinson (1989), Sindiga (1995), Wilson (1997) and Kenya (2002b) identified negative impacts of the trade on the tourism industry. These observations necessitated the development of survey items to evaluate the relationship between sex trade and tourism, the positive impacts of the sex trade, and the negative effects of the sex trade in Malindi (for detailed items, see Table 8.3).

To examine the relationship between tourism and sex trade in Malindi Area, respondents were required to rank their agreement with the 14 items listed in Table 8.3. Table 8.4 indicates that MCSWs in Malindi see a direct linkage between their sex business and the tourism industry. Linkages were found on several items, including: 'Demand for commercial sexual services in Malindi is due to tourism development' (mean = 4.6, SD = 1.1); 'Sex trade in Malindi depends on tourism seasons' (mean = 4.4, SD = 1.0); 'Tourists pay more money for commercial sexual services than non-tourists' (mean = 4.7, SD = 0.6); 'Tourism industry in Malindi creates employment for CSWs' (mean = 4.5, SD = 1.2); 'Sex trade in Malindi leads to an improved personal income' (mean = 4.6, SD = 0.7) and 'Commercial sexual services in Malindi add value to tourist's experience' (mean = 4.3, SD = 1.1). However, sex workers disagreed with the 14th item, which said: 'Tourists avoid the region because of the CSWs' (mean = 1.1, SD = 0.6). These results reveal that although some MCSWs in Malindi did not solely depend on the tourism industry as a source for their clients, they saw the industry as an important component of their market in the area. This might be attributed to the critical role the industry plays in the area's socio-economic well-being (Kenya 1998, 2002; PTLC 1998; Kibicho 2003).

Table 8.3 Survey items

What is your level of agreement with the following statements?
Scale: 1= Strongly disagree; 2= Disagree; 3= Indifferent; 4=Agree; 5=Strongly agree

1. Demand for commercial sexual services in Malindi is due to tourism development.
2. Sex trade in Malindi depends on the tourism seasons.
3. Commercial sexual services in Malindi add value to tourists' experience.
4. Sex trade leads to an improved personal income.
5. Sex trade leads to an increased level of living standard in Malindi.
6. Sex trade improves the image of the Malindi Area as a tourism destination.
7. Most of commercial sexual services are offered within the tourism establishments.
8. Tourists pay more money for commercial sexual services than non-tourists.
9. Tourism industry in Malindi creates employment for commercial sex workers.
10. Commercial sex workers play a major role in the development of tourism in our area.
11. Level of tourism activities affects the choice of area of operation by commercial sex workers.
12. Sex trade leads to the spread of sexually transmitted diseases.
13. Sex trade affects the local economy negatively.
14. Tourists do not like to visit Malindi because of commercial sex workers and their activities.

Chi-square analyses were carried out to establish the relationship between selected demographic characteristics of the respondents ($n = 73$) and condom use by the MCSWs. Tests for association between education level, origin, age, religion and condom use revealed no relationships in all four cases, with $\chi^2_{(4)} = 6.14$ at $P<0.100$ (level of education by condom use), $\chi^2_{(5)} = 5.67$ at $P<0.100$ (origin by condom use), $\chi^2_{(4)} = 4.51$ at $P<0.100$ (age by condom use) and $\chi^2_{(6)} = 5.05$ at $P<0.100$ (religion by condom use). These values indicate that condom use by MCSWs in Malindi Area was not subject to their level of education, origin, age or religion. This observation can be attributed to consistent HIV-AIDS awareness campaigns by Kenya's government, non-governmental organisations and religious institutions (see, for example, Chapters 6 and 7). Furthermore, like their female counterparts (see Chapter 7), all MCSWs singled out HIV-AIDS as their most feared risk while in business. As a result, all male sex workers used condoms when offering vaginal-related intercourse or penetrative sexual services. According to a MCSW in Malindi, 'condom use gives my clients confidence in my services. It also makes us [MCSWs] worry less about possibilities of contracting HIV-AIDS.'

The general view of the sample ($n = 73$) was that there was a direct linkage between male sex trade and tourism in Malindi Area. The MCSWs reported that

Table 8.4 Percentage distribution for the perceived linkage between tourism and the sex trade ($n = 73$)

Items[a]	Mean	SDb	Strongly disagree (%)	Disagree (%)	Indifferent (%)	Agree (%)	Strongly agree (%)
1	4.6	1.1	6	4	10	69	11
2	4.4	1.0	5	11	7	71	6
3	4.3	1.1	0	0	1	93	6
4	4.6	0.7	0	0	0	87	13
5	3.9	0.9	0	1	9	80	10
6	4.2	1.4	1	2	0	90	7
7	3.6	0.9	22	25	3	39	11
8	4.7	0.6	0	4	1	24	71
9	4.5	1.2	6	5	0	42	47
10	4.1	0.8	10	20	6	41	23
11	3.1	1.0	39	33	4	18	6
12	4.0	0.9	8	11	6	50	25
13	3.0	0.7	38	12	0	43	7
14	1.1	0.6	91	9	0	0	0

Note: [a] Items as in Table 9.3; SD = Standard Deviation.

their business was dependent on the tourism industry (mean = 3.5, SD = 0.6). Although difficult to estimate, a large percentage of the income generated by the sex trade was derived from tourism-related activities (see also Kibicho 2004a). On average, the sex workers felt that 65 per cent of their sexual business was in one way or another directly dependent on the industry. A majority of their daily clients, for example, came to Malindi for business, conferences, sporting events and other tourism-related trips (source: field notes).

In addition, from a social perspective, tourism and sex trade share commonalities in meeting the needs of relaxation and fantasy (Ryan and Kinder 1996). Moreover, as noted in Chapter 2, tourism, by definition, is about temporary roles and responsibilities. Sex workers provide such escapes to like-minded tourists (see Chapter 4). Therefore, tourism and sex trade can be said to be subsets of leisure (Harrison 1994; Ryan and Hall 2001; Jago 2003). Further, it is undeniable that, like tourism, male sex trade in Malindi Area has a substantial local socio-economic impact.

Services Offered by Male Sex Workers in Malindi

There were basically three types of commercial sexual services offered by MCSWs in Malindi Area: male–female sexual services, male–male sexual services and pimping services. This section examines these categories of sexual services in detail.

Male–female Sexual Services

Male–female sexual services should not be confused with the common commercial sexual services in the context of sex tourism, where the man is the consumer of a given sexual performance and the woman is the provider of the service. Here, the woman looks for and consumes the service, and thus pays for it (see also Chapter 5). Contributing 56 per cent to the total male sex trade, this was the most popular form of commercial sexual service offered by MCSWs in Malindi Area.

Eighty-seven per cent of the respondents in this category reported that apart from paying for sexual services rendered, clients voluntarily bought presents for them as a sign of appreciation. They often received gifts or Western consumer goods of some value, like laptop computers, cell phones, pagers, iPods, wrist watches, Walkmans, radios, leather jackets, an invitation to accompany the tourist to the next destination or even an air ticket to the tourist's home country. This made it difficult for MCSWs to be precise about the amount of money clients paid for male sexual services.

Ninety-one per cent of the MCSWs offering their services to female clients did not have fixed areas of operation. This is dissimilar to female CSWs, who operated from specific locations (see Chapter 9). It is of importance to note that MCSWs' operation areas have never been determined by the MWA, as they were in the case of female CSWs (see Chapter 9), mainly because the total number of MCSWs is lower than that of their female counterparts. Moreover, there are more seasonal MCSWs than female sex workers (source: MWA's Secretary). The high level of male sex trade seasonality can be attributed to the fact that men in Kenya's patriarchal society have a relatively wider range of alternatives when it comes to 'career' choice in the informal sector compared to female CSWs. Such options include tour guiding, safari selling, souvenir making, taxi driving and hawking of all sorts of paraphernalia. Thus, the level of competition for business among male sex workers is low compared to their female counterparts. Another departure from female CSWs' modes of operation is that MCSWs have comparatively permanent customers who are met at predetermined locations such as hotels, restaurants, nightclubs, striptease clubs and karaoke clubs.

As in the previous studies by Cohen (1982), Dahles (1999) and Rao (2003) dealing with female CSW–male tourist relations, 71 per cent of the MCSWs maintained their relationships through visits, letters and gifts (see Chapters 6 and 9). Eighteen per cent of them had visited their 'clients/friends' in Europe and in the US. Relationships between German female tourists and MCSWs accounted for 52 per cent of all the sex workers who had travelled abroad to meet their partners. On one hand, this can be attributed to the fact that about 49 per cent of the international tourists to Kenya's coastal region are Germans (see Chapter 3) (Kenya 1998). However, this argument contradicts the fact that over 70 per cent of tourists in Malindi Area are Italians (Kenya 1998; PTLC 1998; Kibicho 2003, 2005a, 2006c, 2006d). This contradiction is a fertile area for future male sex trade-focused research.

Anyhow, German female (sex) tourists are reported to be friendlier and more generous with the local men than tourists from other countries (source: discussion with a MCSW in Malindi). Experience has taught the MCSWs that there is a hierarchy of willingness among female (sex) tourists, ranging from Germans ('the most co-operative') to British ('the least co-operative'). To validate the foregoing, Jones (pseudonym), a MCSW in Malindi Area, observed:

> It is much easier to get a German partner .... Germans are very good, [more] understanding, accommodating, outgoing and friendly than other *wazungu* [white people]. They do not care whether you speak German language. They adore the 'African banana' ... they are warm. However, you have to be a performer [be sexually active] as they are like 'empty bottles' which never get filled up. Yeah, they are like empty vessels waiting to be filled with the men's milky' stuff. They are sexually starved.

On the other hand:

> British women are the worst ... they are bullies. Even if you speak the English [language] like Shakespeare, you will end up nowhere with them. Their relationships with us [MCSWs] are like 'hit-and-run' sex affairs. You sleep with one, but when you meet with her the following day, she pretends that she has never seen you. They are too mechanical. They are yet to discover the 'African sex magic' ... they are sexually cold, in fact inactive. Overall, they rarely co-operate with us ....

Thus, nationality plays a crucial role during the selection of the clients, because 'there is a strong link between the country of origin and generosity,' Jones noted. Clients of Italian origin ranked immediately after the Germans on the CSWs' 'generosity scale'. However, they were reported to have a higher tendency to violence, and thus, just like the British, were avoided by the CSWs as much as possible. This avoidance of some clients was also influenced by the prevailing economic needs of the CSW concerned. The rule here seems to be: the higher the economic needs, the less selective the CSW is. In all the cases, however, pecuniary considerations took precedence over sexual ones. Both male and female CSWs in Malindi, for example, often declined to stay with clients who were 'sexually gratifying' but failed to offer 'sufficient' money. They chose their 'suitors' by financial capacity, and calmly rejected those they considered beneath their economic needs. Teddy (fictitious name), a MCSW, explained this

> We are here for business. We need to be paid. It does not matter how good my client is in bed. The bottom line is that my twin daughters are waiting for food

at the end of the day/night. Nevertheless, if you give me good money and at the same time you are sexually fit, then our relationship will flourish …. Thus, my business motto is simple and clear: no pay, no sex! Or, if you like, no good pay, no sex!

Paradoxically, the price elasticity during the bargaining process depended strongly on how desperate the CSW was. If he or she had had no clients that evening, he or she would accept a low rate easily. If he or she had already earned the daily minimum, he or she accepted only clients with a good price. The author noted that CSWs who earned 'enough' money during a particular night started demanding high prices, even for 'less expensive' sexual services like 'suck' (see Table 9.3). Negotiations are extremely difficult in that situation, and give the impression that the CSW actually does not want another client. One informant explained the situation:

> one has to consider whether an additional client on top of a night's minimum earning renders a profit that covers the additional costs in terms of potential physical and/or health risks ….

In such circumstances, the CSW considers such costs can be covered only if the additional client pays extremely well. Thus, from a capitalist point of view, one could argue that CSWs in Malindi have a short-term perspective, only seeking to survive for one day or night, in this case. They do not capitalise on opportunities and accumulate capital.

Eighty-five per cent of the sex workers offering male-female sexual services were willing to quit the trade once they had saved enough money to start a small-scale business, such as a curio/souvenir shop in Malindi town. However, a more lasting, or even permanent, relationship with a white woman was an extremely attractive prospective for 61 per cent of the MCSWs. Like their female counterparts (see Chapter 6), through marriage they hope to escape the poverty and economic insecurity of their present existence (Cohen 1982, 1988a, 1988b; Ryan and Hall 2001; Kibicho 2004b).

From a dependency perspective, MCSWs can be seen as manipulated by well-off female (sex) tourists, who use them to satisfy their quest for the exotic or sexual adventurism – romance tourism (see also Figure 9.2; Chapter 5) (Britton 1982; Burns and Holden 1995; Brenner 2005). However, from the local context, MCSWs use these relationships to accumulate 'symbolic capital'[5] – both financial gains and enhanced social standing (see also the section on 'beach boys' in Chapter 5).

---

5 'Symbolic capital' refers to one's social status and the prestige that accompanies this status.

Male–male Sexual Services

Male sex workers in Malindi also offer male–male sexual services – homosexuality. However, in almost all cases (96 per cent), the local male sex workers assume the 'feminine subject' position, as discussed in Chapter 4 (see also Graburn 1983). The existence of this kind of sexual service notwithstanding, Kenya is a profoundly homophobic society (Mwakisha 1995; Mwangi 1995; Sindiga 1995; 1999; Ndune 1996). Everyone, it seems, has an opinion on the subject. It is a game where the referee is not the only one with a whistle. For instance, commenting on homosexuality (gay relations and lesbianism), Mwangi (1995: 3) warns:

> it is extremely difficult to challenge the conventional wisdom without being accused of condoning what cannot, and must not, be condoned. Homosexuality is a defiance of normality and a challenge to the norms of the society. Taking the risk of legalising such an anti-social sexual (human) behaviour is foolhardy and I would not wish to gamble so much on the wellbeing of our children and the society at large.

The Kenyan lawmaker, as usual, has not been left behind in this moralisation process. However, like other Kenyan laws touching on sex, the provision that punishes homosexuality is riddled with ambiguity and confusion (see also Chapter 2). It is contained in Sections 162 and 165 of the Penal Code (Jackson 1986). According to these sections, homosexuality is one of the three 'Unnatural Offences'. The first section punishes anyone 'who has carnal knowledge of any other person against the order of nature' (Jackson 1986: 69). The expectation of this law is that there is an established manner in which humans are supposed to approach sex, as there is no definition of the expression 'order of nature'. The question is, what would be our measure of naturalness? The last sub-section of the clause punishes anyone who allows 'another person to have carnal knowledge of him or her against the order of nature. *The person accepting to be so known is the one that is hit here. Section 165 of the same Penal Code illegalises ... act of gross indecency with another male ...*' (my emphasis) (Jackson 1986: 69). One risks five years in prison if convicted. However, based on a thorough check at the Central Court Registrar's Office in Nairobi, no one has ever been convicted in Kenya under this law. Why? It is unclear what the law is trying to stop: 'gross indecency'? As opposed to what – 'mild indecency'? This makes the Act difficult to enforce.

However, despite the above strongly worded condemnation, homosexuality is today a reality in Kenya, especially in tourism destination areas, urban centres, learning institutions and prisons (Kibicho 2004b). According to a local elder (82 years of age and a self-proclaimed homosexual) in Malindi, the practice 'was brought to Kenya's coastal region by the slave traders. It is as old as when the Arabs started arriving here ...'.

For the youthful MCSWs, what other people think of them – social acceptance – is not a hindrance to being in the profession. Forty per cent of the MCSWs lying in the 21–30 years age bracket had the notion that what matters is the amount of money one earns, not the way one gets it. A further 86 per cent of the respondents in this category were in the sex trade due to unemployment. All of them were in the 21–30 and 31–40 age brackets. They had attained a minimum of secondary/high school level of education. The remaining 14 per cent were in the business for sexual satisfaction. They enjoyed offering sexual services to male clients. Like their counterparts who specialised in male–female sexual services, they consistently use condoms when offering penetrative sexual services.

Pimping Services in Malindi

A clear difference exists between how pimping services are organised and delivered in Malindi Area and in Western (sex) destinations like the RLDs in Amsterdam (see Chapters 2 and 5). For instance, these services are illegal in Malindi Area, and thus lack any formal organisation. All of the MCSWs offering pimping services operate as brokers or link-persons or middlemen between CSWs and clients (Kibicho 2004b). They serve what Ryan and Hall (2001) call 'more formalised form of prostitution'. In this form of sex trade, CSWs operate through intermediaries to find their clients, in order to avoid social and legal constraints.

All pimps in Malindi make prior contact with CSWs in need of clients for their commercial sexual services. This enables the pimps to know where the CSWs will be during the evening. The pimp then takes up position in a restaurant, hotel, discothèque, nightclub or bar, where he will be based during the night watch. Sometimes they collude with the workers in these establishments, who direct prospective customers to them for a fee.

Ordinarily, no monetary transactions take place between female CSWs and the clients they are serving. The fee is wholly transacted by the pimp with respect to an agreed-upon flat fee from the pimp to the CSW. In all cases, the pimp receives money for commercial sexual transactions on behalf of the female sex worker. This is a strategy aimed at reducing the possibilities of clients refusing to pay the (female) CSWs for sexual services delivered (for more details on how some consumers avoid paying for these services, see Jacques's account in Chapter 7). The pimp receives an average of 30 per cent commission on the total pay. This percentage can either be calculated based on the total day or night's earnings, or for each client served by the CSW. It depends on which is higher from the pimp's point of view. It should be reiterated that the prices for commercial sexual services in Kenya are highly negotiable, thus clients can pay different prices for the same kind of services offered by the same CSW under similar conditions (see also Chapter 5). As far as the CSWs are concerned, pimps play two major roles. First, they look for clients and then link them with the CSW, and second, they provide physical protection to the CSWs from dangerous clients (see also the section 'Mistreatment of Sex Workers by their Clients' in Chapter 7).

Linking CSWs with clients For quality services, pimps ask for some details about the woman a prospective male customer would prefer. Such details include stature, age, complexion, size of butts and busts, social habits, languages, ethnicity, level of education, and most importantly, virginity status. A respondent reported that normally, customers are willing to pay more for sexual services offered by virgin girls. Because of their virginal – and therefore presumably disease-free – status, virgin girls are in high demand in Kenya's sex industry. Thus, almost all pimps struggle on daily basis to include at least one virgin on their list of female CSWs.

To correctly match customers' preferences with the various female CSWs, pimps also require some important details from the ladies available for sexual transactions, including: height (short, tall, medium), complexion (dark, fair, light), size (fat, thin, short, tall), breasts (erect, sagging, tiny, massive, flat), colour (black, white, other), level of education (none, primary, secondary, tertiary, university), languages spoken, ethnicity, marital status (single, divorced, married), with or without children, age and virginity status.

Concerning the last factor, virginity, Maria (fictitious name), a sex worker in Mtwapa (Figure 6.1), reported that local female CSWs had discovered ways to 'regenerate' their virginity. To regain virginity, they wash their private parts with a concoction made from the bark and roots of a particular local tree: 'Although the experience when washing oneself is not very pleasing, the end result is profitable. In any case, the end justifies the means,' Maria concluded. However, she refused to reveal the name of this 'important' tree.

Protection of the CSW To protect CSWs from dangerous clients, sex workers keep in contact with their pimps during the (sexual) service delivery process. If the CSW feels threatened by a client's behaviour, she is supposed to contact the pimp to tell him where she is. Malindi's pimping fraternity has developed a coded language for this purpose (source: MWA's Secretary). For efficient communication, pimps in Kenya's coastal region have at least two working cell phones at any given moment. When a 'danger' is reported, the pimp organises and co-ordinates a rescue mission involving a team of not less than three men, mostly from the pimping fraternity. According to the MWA's secretary, these rescue services are offered free of charge. However, the MWA pays for logistical arrangements like transportation, hiring of guns and other items required.

In return, female CSWs are required to offer a unique extra service to the pimping business. They are required to introduce at least two girls less than 16 years old into the sex trade each year. According to Noor (not his real name), a pimp in Malindi, 'this requirement is necessary to guarantee the future of our businesses'. He added:

> There is a constantly raising demand for young girls.[6] To keep at pace with this demand, we encourage our clients [female CSWs] to either go back to their

---

6 See the discussion on the importance of virginity in sex trade above.

rural homes to 'seduce' their younger sisters, cousins, friends ... or they recruit from around. These girls are never told that they are coming to work in the sex industry. They are instead told of the good jobs and good pay that they will get, which will lead to a better lifestyle. Once they are here, and since in most cases they do not have money to pay for their transport back home, to pay for their accommodation, or to buy their meals, we introduce them into the 'business' [sex trade] ..., which they accept reluctantly at first before they get used to it. We have hundreds of our gals [female CSWs] who entered the trade like that ... and we do not intend to stop new entrants as long as there is demand.

In principle, if a pimp manages to broker a deal exceeding the agreed flat fee, they share it equally with the CSW concerned. This is an excellent pimp–(female) CSW symbiotic model based on a 'give-and-take' principle, with two commercial sexual service providers – pimps and female CSWs – helping one another to achieve their goals. As in the rest of the country, however, pimps control only a small proportion, about 5 per cent, of the sex trade in Malindi Area (see also Kibicho 2004b).

Eighty-seven per cent of the pimps were in this business due to lack of alternative employment opportunities. Eighty per cent of them were in the 21–30 years age bracket. Eighty-nine per cent would quit the profession if given an alternative source of income. The remaining percentage of the pimps enjoyed offering pimping services, thus they would not like to quit the trade. They lay within the 31–40 age group. They were all convinced that sex trade should be legalised: 'We are lobbying for this recognition in Kenya. It is our right ...,' Rajab (not his real name), a MCSW in Malindi Area, reported. According to him, Kenya's Prostitution Act 'should be repealed as it is found on a wholly outdated and thoroughly repugnant moralistic stance based upon rhetoric and dogma rather than a rational, and more ethical philosophy'. Rajab continued:

overwhelming evidence suggests sex trade could not be eradicated and that the main aim should be to reduce as far as possible the harm it can cause. In other words, if policy on sex trade is in future to be pragmatic not moralistic, driven by ethics not dogma, then the current prohibitionist stance will have to be replaced with an evidence-based unified system aimed at minimisation of harms to the society. This logical, rational and consistent approach will inexorably and inevitably lead to the legalisation and regulation of all harmful forms of sex trade – like child sex tourism and human trafficking – in place of the current policy of proscription and haphazard and in fact subjective enforcement. In simple terms, this policy is irrational, illogical, hypocritical and therefore prone to fail ....

Sixty-one per cent of the MCSWs offering pimping services believed that there was a strong linkage between the sex trade and tourism in Malindi Area. For example, 32 per cent of the pimps served only international tourists, who paid more for sexual services than local residents. Leading on from this, it is clear that tourism supports the sex business, as tourists serve as a key segment of the local sex market. Ahmed (pseudonym), a 25-year-old local MCSW, summed it up by observing: 'when the tourism industry is performing well, more money is injected into the local economy which enables the local residents to purchase our services …'. This implies that there is also a local market for pimping services in Malindi Area. This raises a pertinent question: Can these sex workers be categorised as local entrepreneurs? This question forms the central theme of the following pages.

Male Sex Workers as Entrepreneurs

From a classical viewpoint, an entrepreneur is a person who builds and manages an enterprise for the pursuit of profit, in the course of which he innovates and takes risks, as the outcome of an innovation is usually not certain (Boissevain 1974). Entrepreneurs are thus instruments for transforming and improving the economy and society. They are innovators, and indeed decision-makers, pursuing progressive change. Based on the foregoing, therefore, we can define an entrepreneur as an individual drawn from a minority group of low socio-economic status in society, who in an effort to find an alternative avenue of employment consciously decides to undertake an innovative enterprise, assuming risk for the sake of profit (Boissevain 1974; Phillip and Dann 1998; Dahles 1999; Kibicho 2007).

Male sex workers in Malindi are thus eligible to be called entrepreneurs. Socio-economically, they are poor. They have come to Malindi Town to look for employment opportunities, which in most cases are not forthcoming, given the general lack of opportunities even for qualified personnel elsewhere in the national economy (see also Chapters 3 and 6). After trying other forms of income-generating activities as beach boys and hawkers, they have decided to take up the sex trade in order to survive (see Chapters 5 and 6). They are risk-takers. This is because sex trade is a non-routine occupation that involves a strong element of risk, both bodily and regarding the precarious opportunities for success and riches (see also Chapters 6 and 7). Importantly, Malindi's MCSWs, like most small entrepreneurs, attach a lot of value to (the feeling of) independence and freedom, of being able to build and implement one's own ideas. Of course, the principal drawbacks of this position are great vulnerability and lack of alternatives (Boissevain 1974; Dahles and Bras 1999). Moreover, as Dahles (1999: 8) points out: 'the ethos of independence and freedom favours the *laissez faire* economic policy and a deep distrust towards government authorities that are associated with "interfering" regulations and, above all, taxation'.

However, as entrepreneurs, MCSWs are perpetually looking for an opportunity to make a 'larger kill'. This larger-kill syndrome can be explained by the fact that since one of their main resources, the tourist, is accessible only for a limited span of time, many CSWs are pressured to benefit as much and as quickly as possible from the visitors. In other words, they grasp opportunities for profit as they spontaneously arise, gaining in every possible way from the presence of tourists (see also Chapter 4). They do not attempt to build up a stable clientele or a steadily expanding business (Phillip and Dann 1998). They see their 'activities as a set of unrelated exchanges with a wide variety of trading partners and customers that form no overall pattern and build toward no cumulative end' (Dahles 1999: 190). Moreover, they operate under strong competition from their colleagues, as well as tourism seasonality, which do not allow them to establish a stable clientele. Tourism seasonality leads to high market fluctuations and fierce competition among CSWs (see Chapter 6). As a result, the CSWs' business seems to be a combination of a series of cycles in which one oscillates between being ahead of the game and being behind it, between being well-off and being bankrupt. The volatility of these businesses is manifested by the CSWs' spending patterns – they are generous with their friends until broke, when they return to their areas of operations to look for the next clients (see Chapter 5).

Boissevain (1974) distinguishes between two distinct types of resources that are used strategically by entrepreneurs. First-order resources include land, capital, specialised knowledge and other considerations which the entrepreneur controls directly. Second-order resources are strategic contacts with other people who control such resources directly or who have access to people who do. Sociologists call this social capital (see Chapter 9). The importance of social capital in the strategies of the CSWs, in general, underscores the fact that sex workers are not merely economic agents, but most of all social actors (see Chapter 6) (Boissevain 1974; Dahles and Bras 1999). In any case, their economic transactions are also social transactions, as they are more than often embedded in social relations, and not only dictated by impersonal forces (see Figure 9.2).

Entrepreneurs who primarily control first-order resources are called *patrons*, while those who control predominantly second-order resources are known as *brokers*. Normally, patrons manipulate the private ownership of means of production for economic profit. Brokers, on the other hand, act as intermediaries: they put people in touch with each other, directly or indirectly, for a profit (Dahles 1999; Dahles and Bras 1999; Kibicho 2004b). Thus, they bridge gaps in communication between product/service producers and product/service users. Strategic contacts as well as temporal and spatial flexibility to maintain and expand these contacts are important conditions for brokers to operate successfully (see, for example, Dahles 1999; Kibicho 2004b).

Using Boissevain's distinction of entrepreneurs, this study found that the majority of MCSWs (77 per cent) in Malindi are patrons, while 23 per cent are brokers who make a living exclusively by manipulating second-order resources. The first group comprises MCSWs (56 per cent) who offer commercial sexual services to female clients, and 21 per cent who prefer to serve male customers, while the second group of sex workers (23 per cent) operate as pimps. This last category of MCSWs-cum-small-scale entrepreneurs is the most flexible, moving freely within a tourism destination area. Ordinarily, they dispose of a large network and up-to-date information on issues relating to the sex industry in general. More precisely, they have free access to information about tourists, both domestic and international. For instance, they are informed about tourist arrivals, their country of origin; when they will arrive, where they will be staying, the length of their stay in the destination, where they will go next, their activity patterns, their expectations and needs, and their spending power. However, MCSWs refused to reveal who furnished them with these details: 'It will be a contravention of our professional ethics,' the MWA Secretary warned. Subsequently, they are able to match demand and supply in a way that enables them make a profit while at the same time making sure that all the other parties concerned are satisfied.

On the contrary, patrons are more limited in their freedom because a quasi-permanent area of operation ties them down at set times, but nevertheless offers them good connections in the local sex industry. In sum, the patrons depend on brokers for the marketing of their sexual products and/or services, while the latter group depends on the former for their commission. Patronage and brokerage in Kenya's coastal region sex industry thus constitutes a 'safety valve' which allows CSWs to operate in a harmonious and flexible manner.

Although the opportunity to have sex with a woman (female tourist) of different phenotypical and cultural characteristics, without the obligation to marry or at least provide for her, constitutes an enormous stimulus, the woman is an investment in terms of economic and social capital (see Chapters 5 and 9). Male sex workers act as small-scale entrepreneurs – entrepreneurs in the sex/tourism industry. As entrepreneurs, therefore, they have to seize their opportunity under pressing limits of time, as the tourists normally stay in visited destinations only for short periods. In Kenya, they stay for a week to ten days in beach resorts, and two to four-and-a-half days in safari destinations (see, for example, Kibicho 2005a, 2007). Within this short span of time, CSWs try to benefit maximally from these relationships – the larger-kill syndrome discussed above. It is a business of an economical as well as emotional nature (Dahles 1999; Kibicho 2004b). Such liaisons go on for weeks, months, and sometimes years. However, MCSWs break off these relationships immediately the profit drops, eventually entering into another one as soon as an opportunity arises (see Teddy's comments earlier in this chapter). In any case, some of these sex workers have a number of relationships going on simultaneously (see, for example, Chapters 6 and 9).

In other instances, steady MCSWs–female tourist relationships are used as a vehicle to start a business. In such a situation, therefore, the relationship can be said to have a long-term goal. In fact, female tourists are also regarded as potential associates who bring in money to start a business project. The foreign woman provides capital, while the local man becomes a business associate providing access to Kenya's heart-breaking and non-productive bureaucracy, and where necessary, to local networks. As a consequence, becoming a successful businessman as a result of 'mixed marriage' is another (new) dream by a number of MCSWs in Malindi. Foreign female tourists are thus seen as means towards the realisation of these dreams. This is a striking departure from most of the conventional CSWs from other developing countries whose ultimate goal in initiating relationships with the foreign tourist is to acquire a ticket to the Western world (Naibavu and Schutz 1974; Symanski 1981; Cohen 1982, 1988a, 1988b; Gallagher and Laquer

1987; Leheny 1995; Ndune 1996; Oppermann 1998; Cabezas 1999; Dahles 1999; Bauer and McKercher 2003; Kibicho 2004a, 2004b, 2005b; Chapman 2005).

This change in tactics can be explained by the fact that a number of MCSWs in Malindi have visited one or more European countries; in almost all cases, their 'girlfriends' have invited them. For many, whereas they expect to lead a prosperous life in these countries, it turns out to be a total disappointment – different lifestyles, jobs difficult to find, relationship break-up leading to untold boredom and homesickness. They eventually return to Malindi, where they continue searching for a ticket out of poverty. They once again target female tourists as vehicles to make these dreams come true. Thus, we conclude this section by saying that in Malindi Area, and probably in the whole of Kenya, the reasons for entering sex business are changing as the society and wealth of the country change.

Conclusion

This chapter has discussed the relationship between the tourism industry and the development of male sex trade in Malindi Area. The primary intention of the chapter was to explore the MCSWs' roles and to evaluate their dependency levels on the tourism industry when conducting their business. The overall interpretation of the findings reveals a direct perceived linkage between tourism and male sex trade in Malindi. However, sex business in Malindi is a reservoir of hidden unemployment, as well as an overt innovative and enterprising force that is integrated in the local economy. Subsequently, it is true to argue that male sex trade is a form of economic activity, just like tourism. The two activities exist side-by-side in a dual system. This sex (trade)–tourism coexistence necessitates an in-depth scrutiny of how the two industries relate to one another, which introduces us to Chapter 9, 'Tourism's Influences on Sex Workers' Operations in Malindi'.

Chapter 9
Tourism's Influences on the Sex Workers' Operations in Malindi

Introduction

This chapter uses the case study of Malindi Area to analyse the operations of CSWs in Kenya. It is a continuation of the arguments put forward in Chapter 8. Therefore, it reports the findings of a survey conducted among CSWs in the area in the period January–July 2002. As we shall see later, CSWs in Malindi have an established social welfare association, the MWA. From a general viewpoint, this association sets the rules of the sex game in Malindi (see Chapter 2). It structures incentives in human interaction (CSW–CSW or CSW–sex tourist), and thus reduces uncertainty by providing a structure to CSWs' daily life (Figure 1.2). By reducing uncertainties, the MWA makes it possible for 'productive' sexual interactions to take place. It makes it worthwhile for CSWs to enter into commitments and provide some assurance that other (sex) players will uphold their end of the bargain. Within the MWA's 'black box', the production process takes place. However, this process is heavily influenced by production functions outside the 'black box', notably the local tourism industry.

Tourism development that generates net benefits for the local people and protects socio-cultural and environmental aspects of a destination will place restrictions on human activities and challenge the current rapid growth development model. It will strengthen strong moral imperatives, but weaken profit margins (Bachmann 1988; Honey 1999; Gursoy et al. 2001; Alleyne and Boxill 2003). Therefore, relying on 'communal interest' arguments to rally support for small-scale tourism ventures, as mentioned unceasingly throughout this text, risks losing support from the rich multinational tourism corporations when it appears they have little to gain (Britton 1982; Dewailly 1999; Brown and Hall 2000). This is because tourism being a commercial activity, altruism plays second fiddle to profit. None the less, if a tourism destination designs a strategy which makes a stronger ethical argument where all tourism actions are founded on altruism as a central principle, then there is a possibility that tourism development will benefit the poor host communities. At worst, it can be used as a 'policy myth' which encourages awareness, debate and a sense of social purpose at a conceptual level. Anyhow, due care is necessary in order to save the 'altruistic tag' from being misused as a marketing gimmick. Otherwise, like conventional tourism, this form of tourism will be overburdened with ideals it cannot realise.

Leading from the above, therefore, the focus of this chapter is to highlight the implications of tourism-related sex trade for the development of sustainable tourism. It examines the relationship between tourism performance and CSWs' operations in Malindi Area before looking at the development of sustainable tourism. This chapter therefore brings together concepts, theories, arguments and findings to show how sustainable tourism can be used as an antidote to mass tourism in Malindi, Kenya, and probably in Africa at large.

Collection of the Sex workers' Operations Data

As noted earlier, due to the illegality of sex trade in Kenya, there is no official documentation of activities associated with CSWs' operations in Malindi resort (see Chapters 6–8). For this reason, as in Chapter 8, a list of CSWs registered with the MWA was used as the sampling frame.

Using computer-generated random numbers, a sample of 340 CSWs, both male and female, was generated from the MWA's list of 486 members. Equal numbers of potential respondents (170) were approached during the peak season (January–March) and the off-peak season (May–July) in the year 2002. In all, 184 interviews were conducted – 98 in the peak season, and 86 in the off-peak season – with response rates of 58 per cent and 51 per cent respectively, or 54 per cent overall. Given the sensitivity of the subject and the circumstances under which it was undertaken (see 'Setting the Scene'), this response rate is seen as acceptable.

Questionnaires in Kiswahili and English were used as the main survey instrument. They comprised two parts. The first part sought demographic details of the respondents, including age, marital status, level of education, origin and religion. The respondents' profiles are summarised below. The study sample was highly skewed in favour of female CSWs, who constituted 78 per cent of the overall respondents. Sixty-eight percent of the respondents has a secondary school level of education, and their median age was 25–35 years. Eighty per cent of the CSWs came from outside Malindi Area, with 66 per cent of them coming from the Kikuyu community, the largest ethnic grouping in Kenya. Eighty-three per cent of the respondents has been 'pushed' into the sex trade by unemployment, and 8 per cent in the business were married.

The second part of the questionnaire sought information on how the CSWs set their prices, their target clients, their areas of operations, and the general organisation of sex trade in Malindi.

Malindi Welfare Association and Sex Trade

Organisational Structure

Although CSWs in Malindi Area are self-employed individuals, a considerable number have joined the MWA, which contributes to their income security and assists in risk avoidance. Of interest to note is the name of the Association – Malindi Welfare Association. The choice of name for the organisation 'was intended to make it difficult for outsiders to know what the Association is all about or what it stands for as our business [sex trade] is charged with emotions, social prejudices, socio-cultural and political bias,' the MWA's Secretary observed.

The MWA shares some characteristics with other trade unions in Kenya. Such characteristics include having an office, and thus a physical address; and most importantly, establishing insurance against sickness, along with other financial aid. However, it is not formally recognised by Kenya's government due to the illegality of sex trade in general. It is, in fact, an underground organisation.

Female CSWs operating from a common resort have well-knit peer groups, with a high level of 'sisterhood' spirit, often referring to one another as sister. They are ready to psychologically support one another in times of emotional crisis. From a general point of view, therefore, CSWs in Malindi can be distinguished into 'organised' and 'unorganised' categories. The Unorganised are those CSWs who are not registered with MWA. These sex workers are always on the move looking for clients, and have no fixed operation areas, unlike the members of the MWA – the organised CSWs.

The main goal for the MWA is to assist its members in times of financial difficulties. The organisation operates a bank account with a saving system, to which each member must contribute. The MWA normally charges an entry fee of K£200 for new members. If the members do not have ready cash to pay for membership, they are required to nominate two friends from among the MWA members who will pay the registration fee for them. According to the MWA's rules, such a loan must be repaid within six months, at an interest rate of 10 per cent per month for the first three months, and 18 per cent per month thereafter. These high interest rates are intended to motivate the borrowers to refund the money to the lenders within the shortest period possible.

For MWA members with financial difficulties, the MWA provides a loan, with a monthly interest of 0.5 per cent for up to three years. The low interest rates on loans charged by the MWA to its members contrasts with other 'legal' social welfare associations in Kenya which charge at least 12 per cent per month (Kenya 2006). Any member applying for a loan is required to have at least two guarantors, both of whom must have been registered members for at least two years, and who should be regular contributors to the MWA's coffers. The maximum amount of money to be awarded depends on the amount the guarantors are willing to vouchsafe the affected CSW. In case of death of a CSW, the MWA takes full charge of the burial arrangements. Other expenditure which qualifies for financial assistance from the MWA includes house

Table 9.1 HIV-AIDS prevalence by division in Malindi District (as of 2002)

Division	No. of HIV-positive cases	%
Malindi Town	14,703	57.5
Malindi Rural	6,027	23.5
Magarini	3,721	14.5
Marafa	1,158	4.5
Total	25,609	100

Source: Kenya (2002b: 33).

Table 9.2 HIV-AIDS prevalence by industry, 2002

Industry	No. of HIV-positive cases	%
Agriculture	2,876	11.2
Transport	2,425	9.5
Tourism	10,926	42.6
Public service	1,653	6.5
Unemployed	4,651	18.2
Construction	3,078	12.0
Total	25,609	100

Source: Kenya (2002b: 33).

rent – CSWs in Malindi Area usually live alone or with another woman in a rented room – and hospital bills, both for CSWs and their relatives. The MWA's Secretary notes that hospital bills for members currently take the lion's share of the MWA's monthly budget allocation. Related to CSWs' health, 90 per cent of the respondents identified HIV-AIDS as their most feared STD.

To give a general idea of the relationship between the tourism industry and HIV-AIDS prevalence in Malindi District, let us look at the epidemic's distribution by division, then by occupation. The HIV-AIDS distribution by division shows that Malindi Town has the highest prevalence, with 75 per cent of the total (see Table 9.1) Malindi Municipality leads the other divisions in terms of HIV-AIDS prevalence, due to its higher concentration of migrant labour and the tourism industry (Kenya 2002b). The tourism industry attracts CSWs from rural areas and from other parts of the country (see Chapter 8 and Table 8.2). Likewise, the distribution of those who are HIV-positive by occupation indicates that the tourism industry accounts for almost 43 per cent of the total HIV-AIDS cases in Malindi Town. Unemployed people come second, accounting for a 18 per cent of the total HIV-AIDS cases within Malindi Municipality. Importantly, about 21 per cent of those reported to be unemployed work as CSWs, thus they also serve the tourism industry. The prevalence by occupation is shown in Table 9.2.

Sex Trade Operation Zones

The area in the city in which a MWA member operates is determined by the MWA. According to the MWA's Secretary, tourism has made certain locations in Malindi extremely popular with CSWs. Without the social control and discipline imposed by the MWA, major tourist areas would turn into 'battlefields', with hordes of CSWs competing for clients. The Secretary summed up the situation:

> If there are no rules, the game cannot be played. If these rules are less explicitly present, the CSW will have to be more consciously concerned with making his/her own rules, his/her own contact with sexual service consumers.

The MWA determines the maximum number of CSWs that can operate from various zones, depending on the number of tourists who frequent the area, and therefore its business potential. The MWA has accordingly divided Malindi Area into three operation areas: 'high-potential zones', 'potential zones' and 'low-potential zones'. The MWA relies on information from various establishments, notably tourist hotels, to find an approximate number of potential clients in different areas. Surprisingly, the MWA reported that working relationships between the CSWs and different establishments (including security personnel) were generally cordial. It should be noted that police officers are the most disliked government personnel by the CSWs in most sex destinations (see, for example, Cabezas 1999; Kempadoo 1999; Kibicho 2004a).

High-potential zones A designated high-potential zone is characterised by the presence of many hotels, guesthouses, discothèques, restaurants, bars, casinos and shops. Malindi's main street (Kenyatta Street, popularly known as 'K Street') in the neighbourhood of the 'Club 28' discothèque fits this picture, and is the favourite for 71 per cent of the respondents. According to the MWA's Secretary, about 15 per cent of CSWs in Malindi work in the high-potential zone. Sex workers operating from the high-potential zones profile themselves as a separate group. They are more protective and strict about MWA's rules regarding operation zones, which require CSWs to stick to their designated areas of operation. Of course, this is a strategy to shield themselves from possible competition from CSWs working in other zones.

Most CSWs operating in the high-potential zones had been in the sex trade for longer compared to their counterparts in other zones. All of them spoke English fluently, and a number spoke at least rudimentary Italian and German (see, for example, Sophie's story in Chapter 7). They belonged to the high-class sex trade as depicted in Figure 1.1. Their earnings were high as they had a relatively higher revenue retention capacity. This can be explained by the fact that CSWs in this zone dealt directly with their clients. There were no brokers/middlemen between them and their customers. Their level of professionalism was superior to the CSWs operating in the other two zones. This was complemented by higher levels of education. Targeting the elite segment of Malindi's sex market, these CSWs were the cream of the local sex worker community.

The charges of CSWs in the high-potential zones ranged from K£300 to K£600 per service. Sixty-four per cent of sex workers operating within these zones served at least 25 clients per month, and on average, charged K£450 per sexual service. A CSW in these areas therefore earns a net salary of K£11,250 monthly. These earnings are 'tax-free', as sex trade is illegal in Kenya and thus not controlled by the government (see Chapter 2) (Jackson 1986). It merits noting that very few experienced professionals earn these kinds of salaries in Kenya. This of course does not take into account the 'abnormal salaries' for the members of parliament, who receive a tax-free pay package of about K£37,500 a month. It is, however, fair to note that given space and time, Kenya's politicians can easily justify this kind of pay. They are capable of justifying almost anything (risk, uncertainties or sacrifice), and make compromise seem cowardly and criticism treasonous.

Potential zones Potential zones are areas frequented by many tourists, and thus offer the CSWs allocated to them lucrative (sex) business opportunities. Streets adjacent to the entrance to Malindi Marine National Park, Vasco Da Gama Road and the area around The African Curio Market are examples of this type of zone. Potential zones accommodated about 31 per cent of the total CSWs in Malindi (source: MWA's Secretary). Fifty-five per cent of the CSWs in these zones contracted the services of pimps (see also Chapter 8). Celine (not her real name), a CSW who operated along Vasco Da Gama Road, explained:

> pimps link us [CSWs] with local tourists while we pursue international tourists. Some times, they also get us [CSWs] business from *wazungu*. This assures us [CSWs] that we will serve at least a client each day. Of course, they also gain as we pay them a commission. It is a win–win situation.[1]

Compared to CSWs in high-potential zones, sex workers in potential zones are less protective of their territory. They are more permissive towards other CSWs and their colleagues from the high-income-territories. However, they are less accommodative when dealing with CSWs from the low-potential zones. From a general point of view, these CSWs operate in what can be called the 'sex trade buffer zone' – the zone that separates the poverty sex trade from the high-class sex trade. They therefore fall under the tourism sex trade, as detailed in Figure 1.1.

Low-potential zones Low-potential zones, on the other hand, are mainly outside the city centre, in the middle-class residential areas on the outskirts of the town. These zones accommodated the remaining 54 per cent of sex workers in Malindi Area. Ninety-one per cent of them operated as street CSWs. Three-quarters of the street-based sex workers operated independently, while a quarter worked in

---

1   See also Chapter 7.

collaboration with pimps (see Chapter 8). They also delivered sexual services from their residences. This kind of sex trade is known locally as *mama veranda*, or 'women on the balcony'. A female CSW can be found outside her house during the day and in the evening. Balcony women are easier to locate, since their potential clients need only their eyes, and not the local language, to find the right place. The presence of these women on their balconies signifies their availability to serve the next client. The remaining 9 per cent of the CSWs operating in the low-potential-zones were attached to some clubs – karaoke clubs, nightclubs, casinos, discothèques and bars.

The areas called Kisumu Ndogo and The Embassy fall under this description. These zones are characterised by a purely 'local' clientele and near absence of international tourists. According to Zaituni (not her real name), a local female CSW:

> Sex business involving local residents [as clients] offers no opportunities for lucrative offers in terms of tourists gifts, accompanying tourists on excursions, visits to nightclubs, other leisure activities or even a free ticket and invitation to stay in Europe or getting married to a white man. We have to wait until we get promoted to Club 28 ....[2]

These CSWs fall under the poverty sex trade category, as shown in Figure 1.1. They serve all customers willing to pay for their (sexual) services. Compared to the other two zones, prices for specific sexual services are relatively low. CSWs in the low-potential zones are therefore obliged to serve a higher number of clients if they are to earn the same amount of money as their counterparts in the higher-potential zones. On average, a CSW in the low-potential zone serves 53 clients per month. They charge K£115 on average, thus earning about K£6,125 per month.

However, these low-potential zones provide CSWs with regular clients, and thus a low but stable income. Unlike CSWs operating from the high-potential zones, sex workers in these less profitable areas are not keen on the MWA's rules regarding operation zones. All new members of the MWA operate in these zones at first, irrespective of their socio-economic status. However, this is not the case for those CSWs who are in the sex business for pleasure (Chapter 6) – the high-class prostitution on the sex trade spectrum (see Figure 1.1). All the pleasure-seeking CSWs in Malindi Area started in, and still operate from, the high-potential zones. This is possible because in all cases, these CSWs entered the sex trade before applying for MWA membership. As a result, the MWA had no say in their initial areas of operation (source: MWA's Secretary).

Malindi Welfare Association: A Source of Social Capital

Ever since Lyda J. Hanifan, a supervisor of schools in West Virginia US in 1916, introduced the concept of social capital, many researchers have recognised that

---

[2] 'Club 28' is a reference to the high-potential zone.

(social) institutions endowed with this elusive quality tend to function more effectively (see, for example, Jacobs 1984; Mensah 2005). As a result, there is an increasing appreciation of the importance of trust in determining how well organisations (both formal and informal) function. In other words, when an organisation has within it a high degree of trust and co-operation, it is easier for its members to arrive at arrangements of a quasi-contractual nature than if the prevailing culture is one of betrayal and opportunism (Jacobs 1984). In the present context, social capital creates an environment in which STC members tend to make their best efforts to accomplish mutually beneficial ends. While this involves goodwill, fellowship and social intercourse among the members who make up a social unit that would be helpful to the wellbeing of the sex trade in a destination, social capital cannot replace the fundamental benefits of possessing sufficient capital and skilled labour. However, its principles of teamwork and social networking are important for individual survival in a business environment marked by risks and uncertainty as in Kenya's sex profession (see also Chapter 7). To ensure harmonious sex trade operations, the MWA sets prices for various sexual services (see Table 9.3), determines operational zones and settles disputes involving its members.

The monthly fee CSWs pay to the MWA varies between the three zones. Sex workers working in high-potential zones pay K£250, those in potential zones pay K£200, while those operating in low-potential zones pay K£150 (see Table 9.3 for the exchange rate). According to the MWA's Secretary, due to the high numbers of CSWs operating from the low-potential zones, they contribute about 40 per cent of the MWA's income, high-potential zones account for about 33 per cent (due to high pay rates from tourists and the local elites) and potential zones account for the remainder, around 27 per cent.

Although the MWA is in charge of the allocation of areas of operation in Malindi Area, members are first given a chance to choose their preferred zones. If too many CSWs are interested in operating from the same area, some members are placed into a second level of competition. At this stage, the members concerned are chosen randomly by a three-member Business Committee of the MWA. Usually, the MWA advises its members to stick to their assigned areas of operation. According to Celine (mentioned earlier in this section), 'the regularity of operation area provides one with regular clients and, therefore, regular sources of income'. Moreover:

> knowledge of place means that CSWs become familiar with the quirks, prejudices, and demands of hotel owners ... good, bad and indifferent types of potential customers who come to the area ... and most important of all – the policemen who patrol. (Symanski 1981: 76)

This fact has resulted in the establishment of territorial claims in Malindi by CSWs.

Sex workers are accordingly supposed to report those who are not sticking to their designated areas of operation to a five-member Disciplinary Committee for appropriate disciplinary action. Ordinarily, such a breach of operational rules incurs a fine of K£175, which is equivalent to a CSW's full night's earnings (see Table 9.3). This committee, popularly known as the DC, also allocates 'positions' (especially in high-potential zones and potential zones) left vacant by CSWs who quit the trade. Apart from handling disciplinary matters, the DC is also in charge of settling disputes among MWA members. However, disputes between members of MWA and non-members are more difficult to resolve. According to the MWA's Secretary, this is because non-members are not duty-bound to adhere to the MWA's rules. She added:

> in most cases they [non-members] are the ones who provoke our members. Sometimes, they go as far as interfering with on-going CSW–client relationships. This is the worst kind of tricks in a business ....

As mentioned in Chapter 7, in some cases such disputes end in fights, which then degenerate into inter-group conflicts as members of different groups defend their 'sisters'. Sometimes these conflicts evolve into gangster-style fights, when armed thugs are hired by the disgruntled group of CSWs to defend their interests. In response, the attacked group of sex workers hires their own gangsters for a counter-attack. According to the MWA's Secretary:

> the worst inter-group fight was in December 1997. A member of our Association was killed, while the unorganised CSWs lost two sisters [CSWs]. It was the mother of all sex workers' wars in Malindi. The fight was aggravated by the fact that the local tourism industry was in the doldrums.[3] Consequently, competition among CSWs was at its peak. In consequence, sisters [CSWs] started fighting over clients .... The fight attracted the attention of the police, especially due to complaints by the public. In the end, we [CSWs] all suffered – we lost business, some of us were put in prison, and worst of all, our sisters [CSWs] were killed. Today we preach understanding and tolerance .... It is a key component of our mandate.

It is obvious from the above discourse that the presence of tourists forms the crucial criterion for the potentiality of (sex) zones. Sex workers who are fortunate to operate from high-potential zones have organised themselves to protect their position against competition from other sex workers. However, this strategy only works against those CSWs registered with the MWA, as non-members (unorganised CSWs) are not bound to respect the MWA's ethics.

Viewed objectively, the establishment of the MWA is not only aimed at developing business contacts, but also to raise social standing and enhance political

---

3   See also Chapters 3, 6 and 8.

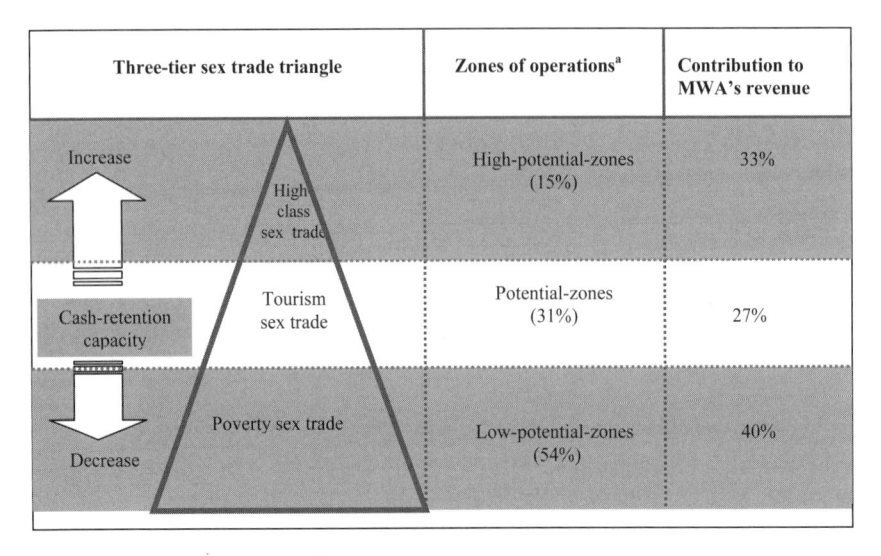

Three-tier sex trade triangle	Zones of operations[a]	Contribution to MWA's revenue
Increase ↑ Cash-retention capacity ↓ Decrease — High class sex trade / Tourism sex trade / Poverty sex trade	High-potential-zones (15%)	33%
	Potential-zones (31%)	27%
	Low-potential-zones (54%)	40%

Figure 9.1   Types of sex trade based on zones of operation in Malindi

Note: [a]Figures in parentheses are percentages in relation to Malindi's sex business.

influence, albeit from underground, which in turn contributes to economic success (see Chapter 6). Such welfare associations – a source of social capital (see also Chapter 8) – are important not only for successful business dealings, but also as an insurance against an uncertain future.

From an academic perspective, the existence of the MWA adds a professional touch to the sex business in Malindi Area. Every profession has four major characteristics: (1) a common body of knowledge, (2) certification of members, (3) self-regulation and (4) a code of ethics. The association manages these four areas. It ensures information exchange and co-operative problem-solving, enables CSWs to update their skills (through seminars), and serves as a voice (albeit in a clandestine manner) for all of its members. Overall, the professionalisation of the sex trade in Malindi is characterised by the adoption (by some CSWs) of a uniform code of ethics designed by the MWA. These rules provide the framework within which CSWs function, and although some may consider the rules onerous, they also provide structure and confidence that things will proceed in a predictable, orderly fashion. They are a guide to interactions between fellow CSWs (CSW–CSW), CSWs and the MWA (CSW–STO) and CSWs and their clients (CSW–ST) (see also Figure 1.2).

However, since what is ethical or unethical is generally open to debate, except in a few clear-cut cases (such as respecting the operation zones), most MWA codes do not tell its members what they should do. Rather, they provide a guide to help sex workers discover the best course of action by themselves. However, based on field observation by the author, there is undeniable failure of the MWA's ethics programmes to achieve the desired outcomes. The major reasons for this are increased competition from unorganised CSWs, lack of clear leadership, increased vulnerability to crime, and decreased public trust.

Based on the potentialities of the various zones according to the MWA's criteria, Malindi's sex trade can be represented as in Figure 9.1. According to Figure 9.1, 54 per cent of the CSWs in Malindi Area operate in the low-potential zones. They lie within the lower section of the three-tier sex trade triangle (see Figure 1.1). Contributing about 40 per cent of the MWA's income, CSWs in this group earn an average of K£6,125 per month. Thirty-one per cent of Malindi's organised CSWs trade their sexual services in the potential zones. Tourists are their main target market. For this reason, the performance of the tourism industry dictates the pace of the sex business in these areas. The remaining 15 per cent of the MWA's members are the elites of Malindi's STC. Lying at the apex of the three-tier sex trade triangle (see Figure 1.1), these sex workers are based in the high-potential zones. They contribute about 33 per cent of the MWA's revenues.

Commercial Sex Workers' Earnings

Sex workers' remuneration takes different forms – everything necessary for a good life, material security, but also professional or social advancement, or simply well-being. They receive expensive gifts, and sometimes 'borrow' money from their clients to buy a house or to start a business, when there is no question of repayment (Kibicho 2004a, 2004b, 2005b). Thus, CSWs' earnings vary, and depend on many factors. However, a sex worker in Malindi Area reported that their income essentially depends on the number of customers they serve and the client's 'purchasing power'. In most cases, the prices are usually fixed in advance based on various factors, such as tourism season (peak or off-peak), duration and type of sexual service to be offered, perceived purchasing power of the customer, and the establishment's charges if the CSW has to rent a room to deliver the sexual services sought by the client. In regard to the last consideration, it is worth noting that in Malindi Area, as in other urban centres in the country, sexual gratification for money can take place any time and virtually anywhere, including, cars, camping vans, public toilets, public gardens, parks, streets, hotels and motels, massage parlours, private homes, bars, dance halls and beaches. Often, it is not easy to ascertain that such activities are taking place except by those 'in the know'.

As a consequence, in interviews it is rarely possible to accurately estimate sex workers' earnings in Malindi, and CSWs are rather reluctant to discuss this issue. Indeed, they prefer to complain about their low pay, and they considered this study as another potential source of income. In addition, due to the trade's unstructured nature and the influences of the high and low tourism seasons, it is difficult to arrive at a clear picture of sex workers' income. However, a price list supplied by the MWA Secretary indicates that the prices charged ranged from K£15 for oral sex to as much as K£325 for a full day's companionship (see Table 9.3).

Table 9.3   Price list for commercial sexual services

Item	Price (K£)[a]
Intercourse	
Short time (= one shot)	25
Three shots	75
Over three shots (= one night)	175
Full day	325
Sex with multiple partners	200
Oral	
Female active	50
Male active	45
Anal	175
Suck	15
Petting	45
Hand/body massage	35
Bondage and discipline	50

Note: [a] US$1 = K£3.8.

All the prices appearing in Table 9.3 are exclusive of food, drinks (bought during the price negotiation period) and the lodging fees (in case sexual services sought by the clients are to be delivered in a rented place). As with other rules and requirements set by the MWA, this price list is only applicable to its members, as the MWA has no control over operations by non-members. The MWA leaves the decision of whether or not to adhere to the stated price list to individual CSWs. In line with the foregoing, 93 per cent of the CSWs do not expressly state their prices, but leave remuneration to the customer's generosity.

As noted earlier, the prices also depend on the persuasive abilities of the CSW, not to mention her capability to create plausible, though often fictitious, stories concerning her needs for money for hospital bills for a relative or other financial problems (see Mercy's story in Chapter 5). Accordingly, CSWs in Malindi Town, in Kenya's coastal region and in the country in general have learnt how to extract money from their clients by appealing to their generosity and their compassion, rather than through outright demands for payment (see also Chapter 4) (Kibicho 2005b). Needless to say, however, the customers' generosity is also enhanced by their level of satisfaction from the sexual encounter. Xing (pseudonym), a Chinese (sex) tourist in Mombasa, described his sexual encounter with a CSW he had spent a night with:

> Surely I have no words to fit the experience. The young lady was sexy, warm, beautiful, affectionate, romantic and sexually very active. She throws you up-

down, left-right ... and keeps on reminding you to push harder. You feel like you are being sucked by a computerised vacuum cleaner – in a well-calculated manner. To be frank with you, I have never had a sexual experience like that before. I was still trembling at check-out ... after a very active night. It was just perfect, no room for rejection. Her performance was priceless. I gave her a small token of appreciation of K£650. Any normal man would appreciate this kind of sexual service. It might not be tonight, but I am definitely trying it again. I have her cellular phone number.

He concluded his description by saying, 'Africano womano is like a concreto machino' ('an African woman is like a concrete mixer/machine'), in an attempt to demonstrate how skilfully and professionally a female CSW moves her waist when providing penetrative sexual services to her clients.

Soft-sell Technique for Sexual Services

In situations where a visiting tourist spends the whole holiday with a CSW, the sex worker is not reimbursed after or before each sexual act. Instead, the payment is packaged in different forms, for instance as support for the education of their siblings or paying for hospital visits for the CSW's mother (see, for example, Mercy's account in Chapter 5). According to Launer (1993), this gives the tourist an illusion that he is with a 'friend' rather than a CSW (see also Chapter 4). Seventy-eight per cent of the CSWs in Malindi Area prefer this soft-sell technique, as it makes it more difficult to interpret the relationship, thus intelligently avoiding being stigmatised by friends and being harassed by the police. According to the MWA's Secretary, incomes from the soft-sell technique range from K£6,000 to K£15,500 per month, with 19 per cent of the CSWs interviewed earning over K£8,725 per month.

Some CSWs have permanent clients/'boyfriends' who normally meet such financial obligations as renting a room, shopping, not to mention the 'pocket money' they send on monthly basis (see Greg's story in Chapter 6). When such a 'boyfriend' returns to his home country, the CSW becomes free to engage herself in full-time sex trade again. Sometimes she looks for another 'boyfriend' who is supposed to meet financial obligations just like the first one. A point worth making is that in some instances, it is the CSW who meticulously schedules when these multiple 'boyfriends' will visit and depart from Kenya, thus ensuring that they will never meet. This intention remains known only to the CSW. In consequence, CSWs can be said to influence, though in a minor way, the temporal distribution of (international) tourists in Kenya. Moreover, female CSWs in Malindi, for instance, have invented a slogan describing their informal polyandry: '*Msikaji, mlipaji na mpenzi*' ('Freeholder, payer and lover'). This means that a female CSW has three men in her life: one she goes out to have fun with (*msikaji*), one who writes the cheques when she needs to buy supplies and clothing and have her hair done (*mlipaji*), and one she is in 'love' with (*mpenzi*).

Amazingly, repeated encounters with the same customer do not necessarily mean the creation of a continuing relationship, nor do they lead to any emotional involvement on the part of the woman. Emma (not her real name), a local female CSW, explained:

> I have been with George [pseudonym] for six years now. Every year he comes to see me here in Malindi. But it does not mean that I love him. I am in love with his wallet. And since the only way to get it is through being with him and having sex with him ... I have no choice but to pretend that I love him. In any case, I am sure he does not love me either, but loves what I carry in between my legs [she smiles then laughs loudly]. Even if we are to get married today, I will marry him because of his money and not his love .... It is true, in the first few years one has some feelings with the clients ... 'made love' with, but these feelings wear out as time goes by .... We call it a 'mark of experience'. Take it from me, 'servicing' these men is a totally depersonalised and unfeeling act.

This revelation by Emma raises the question of who exploits whom in the CSW–tourist relationship (see Figure 1.2). In many instances, CSWs are usually placed in the position of the exploited (see, for example, McLeod 1982; Corbin 1990; Middleton 1993; Harrison 1994); but today some authors are questioning whether the clients, mostly men, actually get what they want (Cohen 1982, 1988a; Bauer and McKercher 2003). Some male clients look for love in a CSW–client relationship, and are disappointed at the commercial approach to sex trade in many encounters (Figure 9.2) (Oppermann 1998; Kempadoo 1999).

Sex Trade: A Double-faced Story

From a positive viewpoint, sex trade as a profession generates opportunities for many Kenyans, young and old alike, with the usual effects of welfare improvement on the side of those involved in the business (see also Figure 9.2; Chapters 4, 5, 6 and 8). This is because the profession has the capacity to enrich and enlarge human attitudes, which goes together with increased income[4] and enhanced trickle-down effects as tourists (both domestic and international) purchase sexual services (Naibavu and Schutz 1974; Cabezas 1999). This is in line with Holloway's (1994) observation that tourism incomes are high in those destinations with many opportunities for tourists to spend. Sex trade expands such spending opportunities,

---

4 The economic impacts of the sex trade can be quite significant. This can be calculated by multiplying the average number of transactions each CSW has per month by the average daily wage bill, times 12 months (53 × K£115 × 12 = K£73,140). Thus, a CSW in Malindi Area operating within the low-potential zones averagely generates K£73,140 per year (see Table 9.3 for exchange rates). Sex workers lying within the high-class sex trade would most probably generate higher incomes than this.

resulting in more tourist dollars being left in the visited destination – reduction of tourism leakage. In addition, it is undeniable that the sex industry can contribute and has contributed to the attractiveness of certain of Kenya's tourism destination areas (Peake 1989; Mwakisha 1995; Sindiga 1995; Ndune 1996; Kibicho 2004a, 2004b, 2005b, 2006c, 2007).

Sex trade also creates good social values, as commercial sexual service providers interact with their customers. Although sex trade may create social problems due to its dynamic capacity to bring about change, thus raising a likelihood of disturbing and upsetting local social arrangements, the resultant social problem can be treated as a necessary social evil in the path of progress, as in Thailand's case (Cohen 1988a, 1988b, 1993; Leheny 1995). It relieves, and to some extent dissolves, some social problems, and constitutes an effective antidote to many others, such as poverty, unemployment and crime (Figure 9.2). Besides obtaining an income in money or kind, providing companionship to tourists may involve the accumulation of social and cultural competence – improving one's learned language proficiency, learning about different cultures, among other things.

Viewed from another angle, sex trade vitiates or destroys wholesome attitudes and undermines the social values of both those involved and the local people around where it is taking place. In other words, the potential economic benefits of the sex industry often come with a bundle of costs which can sometimes be enduring (Figure 9.2) (Harrell-Bond 1978; Cohen 1982, 1988b, 1993; Kelly 1988; Maurer 1991; Ackermann and Filter 1994; Harrison 1994; Hall 1996; Brown 2000). Many commentators believe that nothing but the blocking of sex trade as a destructive social process and then diverting its energy to constructive channels will put an end to the social problems which it generates (see Chapter 2) (Finnegan 1979; Robinson 1989; Findley and Williams 1991; Pickering and Wilkins 1993). It is worth noting that a destructive social process not only breeds problems, but is itself one of the greatest of social problems. This is contrary to a constructive social process, which enriches and enlarges human attitudes, thereby creating social values.

It is, however, important to distinguish between social problems that result from constructive social processes and those that are produced by destructive ones. The former are to be viewed as necessary evils, the latter as preventable. Thus, this theory of social problems begins with the origin of the problems, and ends with their dissolution. Therefore, it clarifies what needs to be done in order to solve any social problem, laying a sound foundation for the development of programmes of social amelioration. It accordingly points the way to both preventive and reconstructive social efforts. Of course, both terms, constructive and destructive, are used here in a relative sense – relative as to time and place. However, it should be noted that the sublimation of the energies of a destructive social process at the beginning of its course into a constructive one calls for the highest prognostic skill and an equal degree of skill in social engineering (see, for example, Fanon 1966; Naibavu and Schutz 1974; Backwesegha 1982; Corbin 1990; Hobson 1990; Stanley 1990; Pickering and Wilkins 1993; Burton 1995; Alleyne and Boxill 2003).

As a result, with a percentage of sex workers in Kenya willing to leave the sex trade in favour of an alternative economic undertaking (see Chapters 6–8), this study presents an important point of departure for the regional and national tourism authorities to start small-scale, locally owned (sustainable) tourism projects to absorb these CSWs and other like-minded Kenyans. In terms of infrastructure and superstructure, such projects are simpler and less expensive than those demanded by mass tourism (see Chapter 3). In addition, locally owned and operated projects will not be enmeshed in the need to conform to the corporate Western identity of the multinational tourism concerns, and thus can have a higher input of local products, materials and labour. In consequence, they will have greater multiplier effects throughout the local economy, and reduce import leakages associated with large-scale, foreign-owned operations. This is considered to be sustainable tourism – at least in this book.

Towards Sustainable Tourism Development

Significant economic benefits can accrue from tourism, and at the same time, socio-cultural and environmental impacts can be minimised through carefully planned and managed development. However, even with the best policies, plans and implementation programmes, it should be recognised that all types of new development, including tourism, bring change. Certain changes are not necessarily undesirable, but can help maintain the vitality of local societies. In any type of destination, tourism is usually only one of the agents of socio-cultural change, albeit a very visible one. Often, it is used as a convenient scapegoat for all undesirable changes that may be taking place in a destination. However, tourism is usually only one element in a complicated pattern of development.

Moral and behavioural changes are certainly occurring in Kenyan society, and indeed in all African societies, but one must be careful not to indulge in romanticism and ethnocentrism by setting descriptions of tourism against some Rousseauesque – or even frightening – idyll of traditional life. In any case, for almost any effect of tourism discovered in one case, one can find a counter-example. For instance, tourism ought to have a symbiotic relationship with the environment; an area often becomes a tourism destination precisely because of its scenic beauty, wildlife and local cultures, among other factors. That attractiveness must survive to lure tourists. Some studies show that tourism indeed preserves wildlife (IUCN 1991), but many others report that tourism destroys the same resource that created it – tourism kills tourism (Smith and Eadington 1992).

In relation to the present subject, sex tourism, a Thai Deputy Health Minister linked tourism to the spread of HIV-AIDS in the country. He said:

> Thailand's profitable tourism industry has been an inhibiting factor in promoting AIDS awareness. Medical officials avoid publicising the appalling AIDS statistics for fear of damaging the country's healthy tourism business …. But

it is long past time for the Government to change Thailand's image as a sexual paradise. [Thailand is no longer 'Thigh-land']. We should promote tourism in a more appropriate way, and campaign more against AIDS. (Robinson 1989: 11)

Commenting on tourism development in Fiji, Naibavu and Schutz (1974) provide an equally strong economic justification for encouraging a relationship between the sex trade and the tourism industry. They argue that sex trade is a fully localised industry, providing income for a large number of unskilled or semi-skilled sex workers (see also Figure 9.2). The industry requires no foreign capital investment, but attracts significant foreign exchange with minimal leakage, further enhancing the trickle-down effects of tourism-related revenue in the local scene (see Chapter 8).

In other words, the linkage between sex trade and the tourism industry in Malindi, in Kenya's coastal region, and in fact the whole country, can also be tailored to tell a similar two-sided story. For example, although CSW–(sex) tourist relationships (Figure 1.2) are said to have negative effects on the local socio-economic setups (for example, through family break-ups and spread of STDs), such encounters can produce a general process of socio-cultural involution. More precisely, the relationship between sex trade and tourism can be depicted along a continuum, with one end representing a positive, mutually rewarding and satisfying experience for all those involved, and the other representing a negative, exploitative and indeed detrimental experience for one or all partners – for example, child sex tourism (Figure 9.2). Pruitt and LaFont's (1985) study in Jamaica, for instance, showed that both female sex tourists and local male CSWs benefit from their relationships. The women found love and companionship, while local men gained 'status', love, and most important of all, financial rewards (Figure 9.2; see also Chapter 5).

However, it should be noted that any oversimplified utilitarian justification for the trade may obscure gender, class, individual and societal problems associated with the linkage between tourism and the sex trade.

Introduction of the Concept of Sustainability to Kenyan Tourism

After the 1970s, it was realised that the development of mass tourism had negative effects on its supporting resources – the human and natural environment (Thomas 1983; WCED 1987; IUCN 1991). In the period, an ecocentric ideology based on the non-consumptive utilisation of the environment and its resources took root in the world (Thomas 1983; Gursoy et al. 2001). From this point of view, mankind, being an element of nature just like flora and fauna, must ensure that the balance of nature is maintained. Mankind must not destroy the natural ecosystem, which is the basis for its own life. Economic development must therefore be considered in relation to the positive and negative consequences for both the human and physical environment.

This ideological shift led to a new school of thought which recognised the environmental values of tourism destination areas. As a result, an eco-sensitive

196 Sex Tourism in Africa

Female sex tourist
- Find love while on holiday
- Find romance while on holiday
- Find companionship from local CSWs
- Enhanced tourist experience
- Cheaper commercial sexual services
- Cultural involution
- Reaffirms their feminism
- Opportunity to penetrate male domain
- Domination of the CSW
- Self-actualisation
- Opportunity to accrue socio-cultural competencies (e.g. language proficiency)
- Opportunity to test stereotypic beliefs of 'Others'
- A chance for social networking

Male sex tourist
- Find love while on holiday
- Find romance while on holiday
- Find companionship from local CSWs
- Enhanced tourist experience
- Cheaper commercial sexual services
- Cultural involution
- Domination of the CSW
- Self-actualisation
- Opportunity to accrue socio-cultural competencies (e.g. language proficiency)
- Opportunity to test stereotypic beliefs of 'Others'
- A chance for social networking

Positive end

Sex worker
- Find love from the tourist
- Find companionship from (sex) tourists
- Feeling of self-actualisation
- Cultural involution
- Domination of the tourist
- Opportunity to learn new skills
- Opportunity to visit foreign countries
- Access the tourist world
- Accumulation of new competencies
- Enhanced personal income - welfare improvement
- Family cohesion as CSWs are able to financially assist their families
- Opportunity to accrue socio-cultural competencies (e.g. language proficiency)
- Self-actualisation
- A chance for social networking

Local economy
- Creation of job opportunities (especially for un/semi-skilled residents)
- Increased spending opportunities for tourist
- Enhanced family earnings resulting in reduction of poverty levels
- Improved local economy (CSW-tourist businesses partnership)
- Enhanced trickle-down of tourism incomes
- Enhanced multiplier-effect
- No foreign investment capital required
- A challenge to the local authorities to design new ways to engage CSWs
- Diversification of income generating activities
- Crime reduction crime (residents have a source of alternative income)
- Improved destination image (especially to the sex tourists)

--- Positive-negative interface ---

Sex worker
- Affected by the spread of STDs
- Increased alcoholism as CSWs adopt tourists' social norms
- Increased drug use to deal with exhaustion
- Reduced self-esteem
- Risk of physical attacks by the clients
- Risks of vaginal/anal tearing
- Reduced level of honesty as CSWs create fictitious stories in order to earn money
- Increased CSWs-law enforcers animosity
- Increased non-marital childbearing
- Destruction of social values
- Increased CSWs' alienation due to social stigmatisation
- Confirms stereotypic beliefs of the CSWs by the (sex) tourists
- Continuation of the colonial *master-servant* relationship between locals and tourists

Local economy
- Weakened human resource due to STDs infections (especially HIV-AIDS)
- Negative destination image (especially by non-sex tourists)
- Erosion of local traditions and socio-cultural set-ups
- Increased child labour due to child sex tourism
- Increased school dropouts
- Increased demand for resources to fight against sex trade
- Increased demand for resources to fight against the spread of STDs infections
- Reduced local people's sense if identity
- Reduced local residents' sense of self-worth
- Continuation of local economy's dependency on the tourist generating areas

Female sex tourist
- Affected by the spread of STDs
- Loss of personal income due to purchase of commercial sexual services
- Affected when relationships with CSWs break-up

Male sex tourist
- Affected by the spread of STDs
- Loss of personal earnings due to purchase of commercial sexual services
- Affected when relationships with CSWs break-up

Negative end

Figure 9.2   Positive–negative continuum of tourism-related sex trade

wave developed which resulted in the rejection of mass tourism by tourist host communities, by eco-tourists, and by conservationists (Chapter 3). This saw the birth of sustainable or alternative tourism development (WCED 1987; IUCN 1991; WTO 2004). One can thus say that sustainable tourism was a result of market demands. The desire for adventure and the unknown is a crucial element in this kind of tourism. It is characterised by a demand for non-specialised services and an intensity of interpersonal relations during the service delivery process. Thus, the interactions between the client and the tourist product/service provider are numerous, as the visitors are well informed, and thus more demanding. In consequence, unlike mass tourism, sustainable tourism does not necessarily insist on the minimisation of costs, but rather assurances about the quality of the service(s) delivered and the varieties of tourist products.

This change in paradigm was integrated into tourism policymaking and implementation in many countries (Brenner 2005). However, this new concept of sustainable tourism presents two major challenges. First, it is difficult to identify quantifiable and applicable indicators to evaluate the impacts and monitor tourism development (Burns and Holden 1995; WTO 2004; Brenner 2005). Second, local people lack capital, entrepreneurial skills, necessary experience and access to tourism markets. Questions also arise about what exactly needs to be conserved, with what objectives, at what cost, and for whose benefit. What are the conditions for sustainable tourism development? Who determines the meaning of the concept of sustainability? Who decides how it should be carried out and evaluated? What criteria and indicators should be used in its evaluation? For this reason, there are as many operational definitions as those who are interested in defining this concept. Ecologists use scientific rigor to support their findings and proposals; developers refer to an art or to a technique; politicians use compromise, while referring to governmental ethics. How can one synthesise these different approaches, to unify these differing 'languages', harmonise this diversity of modes of operations? When we talk about 'sustainability', what exactly do we mean? In brief, the concept of sustainability is powerful in theory, but troubled in practice. Nevertheless, answers to these questions can make operationalisation of the concept of sustainable tourism possible, thus finding a balance between local development and socio-cultural and environmental conservation, and at the same time ensuring tourist satisfaction (Kibicho 2005c).

In this book, 'sustainable tourism' is used to denote a form of tourism that meets the needs of present tourists and host communities while protecting and enhancing opportunities for the future (WCED 1987; WTO 2004). This section restricts its focus to socio-cultural and economic sustainability in tourism, although it is recognised that the concept of sustainability also encompasses ecological dimensions (WCED 1987; IUCN 1991).

Notwithstanding these difficulties, socio-environmental dimensions were added to the hitherto economic approach to Kenya's tourism development in 1989 (see Chapter 3). Thus, according to the government's Sessional Paper No.

23 of 1989 on sustainable tourism development, the objectives of this new form of tourism are to:

1. improve the living conditions of local communities;
2. enhance both material and political participation of local people;
3. prevent environmental degradation or loss of biodiversity. (KWS 1990)

Further, forward and backward linkages with other economic sectors are to be strengthened. However, the extent to which this attractive-sounding theoretical idealism was met by reality remains debatable. With all this government rhetoric, for example, the national tourism policy continued to encourage the development of large-scale mass tourism projects (see Chapter 3). These policies precluded local community participation due to the obvious capital-intensity associated with such tourism projects. Many local people were only involved in the informal sector of the tourism industry, such as selling souvenirs and offering sexual services to tourists.

This situation was compounded by the fact that overseas travel agents designed fast-profit-making tourism packages – all-inclusive tour packages. In these travel arrangements, a tourist pays a travel agent for all the travel components, such as air tickets, food, accommodation and recreational activities, while at the destination. The concept is premised on the assumption of high occupancy rates and minimal contact outside the hotels, thus limiting encounters between tourists and members of the local community. Such establishments thus limit the capacity of the tourism industry to spread benefits outside the environment controlled by the all-inclusive hotels (Alleyne and Boxill 2003). It is 'an intelligent way of enhancing the already high leakage of tourism receipts to overseas companies' (Kibicho 2003: 38).

All-inclusive tours make it possible for an excessive portion of the revenues obtained through tourism to leak away from the local economy (Kenya 2000; Kibicho 2003; WTO 2004). While certain imports cannot be avoided, reducing leakages is important to ensure sustainable growth of local economies. Tourism leakage manifests itself in three general ways: external, internal and invisible. External leakage is represented by earnings which accrue to foreign investors financing tourism infrastructures and superstructures, through repatriated profits and amortisation of external debts. Internal leakage, on the other hand, arises from tourism through imports of tourist equipment and/or services that are paid and accounted for domestically. Lastly, invisible leakage is the opportunity costs which can exert cumulative and significant effects. A major source of invisible leakage arises from the non-sustainability of tourism-related socio-cultural and environmental assets due to an ill-planned and poorly managed tourism industry (see also WTO 2004).

This external control of Kenya's tourism components resulted in meagre trickle-down effects reaching the local communities hosting the tourists. Eventually, the local people felt sidelined from tourism's mainstream activities. Determined to win a share of the tourist dollar, some of them get involved in illegal undertakings such as drug trafficking and trading in sex (see Chapters 3, 5, 6, 7 and 8).

Local Community Participation as a Form of Social Equity in Sustainable Tourism

Accepting tourism as a tool for economic development often means accepting cultural change (WTO 2004), because socio-cultural and economic impacts on a host community are inextricably connected. There may be beneficial synergies or even inverse relationships among the three impact areas. However, opinions about what is positive or negative may differ among local community groups or individuals. These differences can be smothered through a series of actions aimed at continuous awareness-building and engagement of local people (Kibicho 2003). These actions evoke a sense of responsibility and generate a set of common goals which are a prerequisite for sustainable tourism development (Cazes 1992; Dewailly 1999; Honey 1999; Brown and Hall 2000). Further, formulation of a set of community goals through consensus empowers local residents to make choices that suit their common aspirations (see, for example, Britton 1992; Burns and Holden 1995; Walsh et al. 2001). This kind of social equity consideration suggests a greater voice for the host community in negotiations, and a greater stake in the income generated. Ironically, however, the very effort to include all stakeholders in the planning process can be an important obstacle to achieving a balanced form of development, as the power of interested parties is rarely evenly matched (Lickorish et al. 1991; Hall 1994; Cuvelier 1998; Honey 1999). In other cases, elite internal communities have sometimes developed, often to the detriment of the indigenous residents (see Chapter 3). To avoid this bottleneck, this book suggests that the best way to approach an overarching goal of sustainability is to ensure a successful marriage of top-down government/elite agendas with bottom-up/community-inspired objectives. Thus, tourism destination managers should steer clear of the prevailing perception that local residents lack the expertise to make informed decisions.

A practical application of the social equity concept within sustainable tourism development occurs when the host communities who are affected by decisions participate in making those decisions. This gives local stakeholders more active involvement in the development process itself (Walsh et al. 2001). Therefore, sustainable tourism development is necessarily a participatory process (WCED 1987; Smith and Eadington 1992; Dahles 1999; Honey 1999; Walsh et al. 2001; Kibicho 2003, 2005c; Smith and Duffy 2003; WTO 2004). While the impetus may come from the government, from the tourism organisations or from the tourist host community itself, early involvement of all those interested in the industry should be considered essential. The complexity of the stakeholders, their interests and relationships at the local level cannot be underestimated. None the less, the importance of local people's participation in the tourism development cannot be gainsaid. Their knowledge on issues such as local use of resources, key traditions and the values they hold most important regarding the destination is invaluable. In addition, they have clear ideas regarding the current tourism situation, and strong opinions on what is likely to be acceptable in the future. As a result, this kind of participation helps in determining the tourism assets that are critical to their needs and expectations, as well as those of visitors. In addition, they will help in responding to pertinent questions like whether they are sensitive to changing tourists' demands, or whether local residents' values are sensitive to the changes associated with tourism development.

It should be noted that local people's satisfaction with tourism is essential for its sustainability. Actions by all interested parties to maintain a positive relationship between the local people and tourists can prevent community hostility and negative effects in general. This brings us to the perceived linkage between tourism development and sex trade growth in Kenya's leading destination areas (see Chapters 5–8). In Malindi, for example, the tourism industry has been categorised as a social evil (Peake 1989). It should be noted that the Islamic religion has a strong influence on the socio-cultural structure in Malindi Area (Bachmann 1988; Peake 1989; Kenya 2000, 2001, 2002b; Kibicho 2003). Therefore, any activity, which happens to contravene Islamic principles is strongly resented. The tourism industry happens to be grouped together with those industries with unacceptable activities by some local people (Bachmann 1988; Peake 1989; Ndune 1996). The main reason behind this resentment, as Kibicho (2003: 37) notes, is that 'apart from lowering moral standards in Malindi, tourism is one of the most important contributory variables in the spread of prostitution, alcoholism, drug-taking and crime'. Such 'vices', it is believed, are anathema to the local people's religious beliefs and practices.

A tourism destination can have all of the ingredients for success – superb infrastructure and superstructure. However, if there are no qualified local people to provide tourist services and operate facilities, economic sustainability of tourism at the destination level will not be achieved (WTO 2004). The key component of employment that affects this sustainability includes skills acquired, training levels, quality of jobs, turnover, pay levels and tourism seasonality. Due to low levels of training, Kenya's CSWs would probably otherwise work in low-paid positions in the tourism industry. Coupled with problems associated with tourism cycles (see Chapters 3 and 6), CSWs would most likely earn an inadequate income in relation to the cost of living and maintenance of a decent quality of life. This might force the CSWs to work on part-time basis in the sex industry to supplement their earnings (see Chapter 6). In the worst instances, CSWs might abandon the new jobs, opting to go back to trading in sex (see Fatima's and Zainabu's comments in Chapters 2 and 4 respectively). Thus, the tourist host community should be interested in knowing the number of new jobs, who is to get these positions, and how many people are leaving their traditional activities to participate in tourism-related activities (see Chapter 8). Small-scale locally controlled tourism will contribute to the socio-economic well-being of the local people. It will not only provide paid employment, but will also provide greater stability and more opportunity to diversify income-generating activities. Such diversification creates a buffer against external threats whereby shortfalls in one economic sector (for example, agriculture due to drought) can be compensated by other activities (such as tourism). Time invested in tourism versus tourism earnings determines the effectiveness and efficiency of local efforts in producing tourism-desired profit.

Poverty Compromises Tourism Sustainability

A debate on sustainable tourism in Africa can never be conclusive if it fails to examine the implications of poverty for the industry's long-term sustainability. To begin with, it is important to stress the fact that poverty on the African scale is more than an individual phenomenon (see Chapter 1). It is also a social and political one, tightly interlocking with the workings of economies and societies in a multitude of ways. Consequently, abject poverty drives local people to engage in activities capable of compromising the socio-cultural, economic and environmental sustainability of a tourism destination (Honey 1999; Kibicho 2003; WTO 2004). At the margin of existence, concern for security inhibits the adoption of alternative, potentially advantageous and more sustainable socio-economic undertakings like small-scale businesses, thereby reducing their own growth potentials (see Chapters 5 and 8).

A common way to cope with a shortfall in income from agriculture and other self-employment activities in Africa is to search for waged employment (Kenya 1991, 1996, 2004, 2006; Gitonga and Anyangu 2008). While such a strategy might work when dealing with an occasional shortfall in earnings, it can largely be ineffective given the far-reaching effects of abject poverty in a non-performing economy compounded by a large number of rural males searching for jobs at the same time. When wives and children left at home fail to receive any wage income, they themselves leave the village in search of food relief. This sequence of events results in poverty of dramatic proportions. In some cases, poverty makes families vulnerable to accepting food handouts and money from wealthy adults or international tourists. While some make genuine offers, others donate in order to get close to the female family members. Many families ignore sexual relationships between the donor and one of their offspring as a means of ensuring their family's future survival. In fact, some families coerce their daughters to befriend 'big men' or international tourists (Gitonga and Anyangu 2008). In other instances, poverty leads many families to be (physically) separated, as some members venture into selling sex in urban areas as an exit route from this situation (see also Chapters 1 and 6). Whatever the case, the families concerned are constantly in denial about the source of income from their offspring, preferring to believe that they are working in a hotel, a safari company, a restaurant or some other tourism-related enterprise.

Failure to address growing poverty in Africa will increase incidences of violence and crime, which will eventually imperil the development of a viable tourism industry. As mentioned later in this section, any real or perceived risks determine whether tourists return, recommend the destination to others, or conversely, advise others to stay away (Oppermann et al. 1998; WTO 2004). Thus, tourism destination managers should aim to reduce the level of public insecurity in their areas. In an attempt to remedy some of these security problems, some destination areas in Kenya are coming up with community-based tourism projects (KWS 1995, 1997; Kibicho 2003, 2005c, 2007). Such small-scale locally owned projects tend to bring greater benefits to the larger society, because local people outside the mainstream tourism industry often provide additional products and services to tourists (see also Chapter 5) (Alleyne and Boxill 2003; Kibicho 2003, 2004a, 2004b). Further, these forms of tourism projects are less disruptive to local socio-cultural set-ups, are associated with smaller leakages, and are more likely to fit in with indigenous activities and land use patterns (see, for example, Smith and Duffy 2003). Leading on from the above, it is clear that alleviation of poverty levels in Africa is not a luxury decision – the domain of 'big-hearted' pop stars or 'enlightened' bureaucrats; it is a prerequisite for the overall development of tourism in the continent (WTO 2004). Thus, all members of the tourist host communities, including CSWs, should be given economic opportunities, be socio-politically empowered and be provided with security (see also Chapter 6).

Tourist Insecurity and Destination Sustainability

Tourist insecurity threatens the sustainability of a tourism destination. A single man-made incident can cause far-reaching effects as tourists cancel their trips while tour operators redirect their tours to alternative destinations (see Chapter 6) (PTLC 1998; Kibicho 2003). The impacts of such events may last for months or even years after conditions change and order is restored. This is because recovery from damage to a positive image can take a long time, even if the destination is quick to respond to actual or perceived insecurity. A point worth making is that perception of security is at times in the eye of the beholder. Some tourists, for instance, may feel uneasy in a milieu where they are besieged by CSWs or shouted at by beach boys/operators (see Chapter 5). Others, however, may relish the experience and consider them to just be part of the tourism experience, and thus perceive no threats.

Thus, it is not only necessary to note and deal with the objective incidence of insecurity problems, but also to try to understand the level of unease perceived by visitors. In addition, tourists who are victims of crimes frequently share their nasty experiences with others, often in public forums such as online chatrooms or newspaper complaint columns. This will affect decisions by potential tourists on whether to visit a destination or not. If they decide to shun the destination, then this hampers the transfer of economic benefits of tourism to the local economy. Thus, maintenance of public security is a crucial factor in promoting a favourable image of a tourism destination. Therefore, one may say that sustaining tourism destinations involves ensuring that appropriate images of that place are established and are refined to suit evolving markets.

Sexuality and Tourism Destination Marketing

Marketing is a key element in the success of tourism destinations. The creation and maintenance of markets for a destination contributes to its sustainability. Further, socio-cultural factors are important elements of an enhanced quality of tourist experience. Thus, destination marketers seek the degree to which there is a market niche for environmentally sound and socially responsible products and practices. In other words, achieving tourism destination goals depends on determining the needs and wants of target markets, delivering the desired satisfactions more effectively and efficiently than competitors, while at the same time protecting the resource base of the industry.

Tourism, like other commodities, is packaged for exchange by advertising, much of which appeals to people's deepest wants, desires and fantasies (often sexual), and is anchored in a dynamic of image construction, and indeed manipulation (Watson and Kopashersky 1996; Walsh et al. 2001). Thus, destination images are used to instil beliefs, ideas and impressions of a place. The sex that is sold by tourism destination marketers not only consists of sexual services, but also film, theatre and other forms of performance, artefacts, paintings and other printed representations which have explicit sexual and erotic components (Baillie 1980; Cazes 1992; Burton 1995; Gould 1995; Bauer and McKercher 2003). It also includes sexual referents and illusions employed to 'package' non-sexual tourism services.

The tourism industry in general is dependent on the creation of fantasy through imagery, its advertisers being described as 'dream packagers' and 'purveyors of escape' (Burton 1995; Gould 1995; Oppermann 1998; Oppermann et al. 1998; Ryan and Kinder 1996; Walsh et al. 2001). In most cases, tourism promotion focuses on the more licentious attributes of the tourist, and highlights the erotic dimensions of the destination. It promises would-be vacationers more than sun, sea and sand; they are also offered the fourth 's' – sex. According to Baillie (1980: 20):

> resorts are advertised under the labels of hedonism, ecstacism, and edenism ...
> One of the most successful advertising campaigns actually failed to mention the location of the resort – the selling of the holiday experience itself and not the destination was the important factor.

In consequence, certain tourism destinations are so closely associated with sex tourism that the mere mention of such destinations raises assumptions that male tourists are visiting for one reason – consumption of sex (Leheny 1995). Such countries include Cambodia, Thailand and Vietnam. By the same token, some other destinations are recognised as venues where female tourists fulfil their sexual fantasies. This includes countries like Jamaica, The Gambia and now Kenya (Pruitt and LaFont 1995; Afrol 2003; Kibicho 2004b).

Nevertheless, it is worth noting that this usage of sex does not necessarily transform female or male sexuality into a product, for two major reasons. First, sexuality is generally a persuasive motivator in consumer behaviour (Oppermann 1998; Ryan and Kinder 1996; Kibicho 2004b). Sex is thus used to sell everything from soup to nuts. Sex both sells and is sold by these marketing strategies. This is called *product sexualisation*. Product sexualisation involves the act of presenting consumer/tourist products and services in such a manner that they are easily identifiable with some specific sexually suggestive moods. As a result, Gould (1995: 396) concludes:

> just about anything may become a conditioned sexual stimulus when paired with an unconditioned one. Thus, a product becomes sexually conditioned when it is associated with an unconditioned sexual stimulus, such as an attractive model in an advertisement or a sexually attractive person in an actual consumption situation.

Such a sexualisation approach may be used for positioning tourist products and services among competitors. Second, travel can be compared to the phenomenon of love in terms of its varieties and approaches – a process in which the traveller abandons the familiarity of home for exciting new amorous experiences elsewhere (Watson and Kopashersky 1996; Oppermann et al. 1998).

The emphasis in both cases is that what you see is what you get, indicating the attraction of sexual fantasy. The dominance of the female body would suggest that potential tourists or targeted consumers are primarily male. However, this hypothesis is not backed by any statistical evidence. The *female fantasy hypothesis* argues that photographs of female bodies encourage women to live out their fantasies to be as beautiful as the women appearing in the advertisement (Baillie 1980; Gould 1995). Contrary to this hypothesis, Oppermann and his co-researchers note: 'women are more attracted to pictures of couples and ... identify female body shots in advertising as marketing methods to objectify and exploit women' (1998: 23). It tends to represent the woman as a sensual, sexually available and subservient (Gould 1995). This is probably best represented by Singapore Airlines' ongoing promotion which reads: 'Singapore Girl – a Great Way to Fly' (Kibicho 2007).

For these reasons, tourism destination marketers should be careful in their usage of sexual imagery and sexual innuendo in order to save their destinations from being 'black-listed' by those who define the immoral landscape – the moralists. One example is the withdrawal of an advertisement by Austrian Airlines in mid-1993 (Kohm and Selwood 1998). The advertisement for flights to Thailand featuring a youthful-looking girl was placed in Germany. Its withdrawal was necessitated by constant objections that the 'provocative' girl on the advertisement moralised child sex tourism, and would thus eventually encourage under-age sex (Kohm and Selwood 1998; Oppermann et al. 1998). This implies that although sexual content usually enhances the attention directed towards a specific advertisement, it does not necessarily translate into positive feelings about such publicity or the product being promoted (Watson and Kopashersky 1996). However, one should be reminded that sexual provocation depends on the individual's perception and frame of reference, which varies depending on age, education, gender and culture, among other factors.

That said, tourism marketing creates expectations about what is to be found at a visited destination. Consequently, a concerted effort is necessary for destinations to offer products and services that maintain or improve tourists' and host communities' expectations, which will guarantee these areas' sustainability. Thus, all members of the local tourism system should be involved in the marketing process. This will result in creating an acceptable destination image (Gould 1995; Walsh et al. 2001). In this way, tourism destinations will be able to recognise and respond to the needs and interests of local residents balanced against those of the tourists, eventually enhancing local people's sense of identity and self-worth while at the same time providing extra income.

Leading from the foregoing, we close this section by pointing out that where tourism market segments fit the host communities' needs and aspirations, and when tourist products are developed with local people's participation, then unique and indeed vital socio-cultural aspects of a tourism destination will be sustained (Gould 1995; Cuvelier 1998; Honey 1999; Walsh et al. 2001; ECPAT 2002; Kibicho 2003; WTO 2004).

Conclusion

This chapter has revealed that some CSWs in Malindi organise their profession through a welfare association, the MWA. Although Kenya's government does not recognise the MWA, it plays an important role in the social and professional life of its members. In addition, this chapter has found that tourism has affected the lives of at least a fraction of CSWs in Malindi Area, who benefit from tourism in terms of an increase in income. The emergence of tourism in Malindi, for instance, has created opportunities to earn considerably more money from tourists. It is therefore safe to say that tourism enhances the economic capital of those CSWs who get involved with tourists. Further, the chapter has shown that tourism locations are defined as 'high-income territories', and that access to these sites by CSWs is partially controlled by the MWA. This effectively permits the industry itself to exercise some discretionary control over what might become a nuisance to some tourists.

It is apparent that to achieve an integrated approach to the sustainable tourism development in Malindi, in Kenya, and indeed in Africa, all levels of government and the private sector must closely co-ordinate their activities to ensure that the values that support tourism, local people, are not lost through exploitative and indeed unsustainable development. Enhancement of sex workers' skills can be an important strategy towards the integration of the many CSWs willing to quit the sex trade into the tourism mainstream. Such a programme should adopt a systemic approach, where there is recognition that a problem affecting one element of the system will have spill-over effects on every other element of the system (see Chapter 1). In any case, a community is rarely sustained by tourism alone, thus tourism needs to be considered in conjunction with other socio-economic factors.

In brief, the general prosperity of a (desti)nation depends on the well-being of its people. Furthermore, today's complex tourism industry environment requires destinations to be better prepared to anticipate and be pro-active towards internal and external changes. Successful achievement of a tourism destination's objectives depends on the ability of the management to respond to opportunities in an ever-changing environment. Sustainable tourism development promises all this. It ensures that all the decisions and actions used to formulate and implement strategies are designed to achieve the destination's socio-economic as well as environmental objectives.

General Conclusion

It is useless to tell a river to stop running;
the best thing is to learn
how to swim in the direction it is flowing.
Kiswahili proverb

This final chapter is a disquisition of the conclusions drawn from this study. Certainly, such conclusions cannot be presented without reiterating some of the pertinent arguments and issues presented earlier in the book. For this reason, this chapter contains fragments of a summary, but only where relevant to elucidate the conclusions. It is composed of four critical sections: 'Contextualisation of the Sex Trade', 'The Marriage between Tourism and Sex Trade', 'The Complexity of the Tourism-oriented Sex Trade' and 'From Sex Trade to Small-scale Tourism Projects'.

Contextualisation of the Sex Trade

It is clear from this study that sex trade in Kenya, in Africa, and probably in other developing countries, need to be contextualised within the gendered socio-cultural, economic, historical and political backdrops (Figure 3.4). The inter-relationship between socio-cultural aspects of the CSWs and the wider politico-economic contexts needs to be examined in order to develop practical policies to address issues related to sex trade and tourism development in Kenya. Such policies should recognise the fact that politico-economic structures mediate socio-cultural practices in many societies. The law, the health system, welfare and benefit systems, and the media, which are all involved in the social reproduction and constitution of society, are instrumental in mediating sex trade as an economic undertaking (Figure 1.2). Knowledge about these agencies is central to understanding the circumstances, experiences and needs of the CSWs. They help to mediate the actions and attitudes of individual CSWs through a mixture of service provision. For instance, this book has revealed that the most significant societal factor that pushes Kenyans (both men and women) into sex trade is poverty.

It must be reiterated that quite a number of destinations with thriving sex tourism industries are countries that suffer from widespread poverty. These poverty levels result from turbulent political situations and unstable economies. Poverty levels often correlate with illiteracy, limited employment opportunities and bleak financial circumstances for families (Figure 3.4). Children from these families become easy targets to human traffickers. They are lured away from broken homes by recruiters

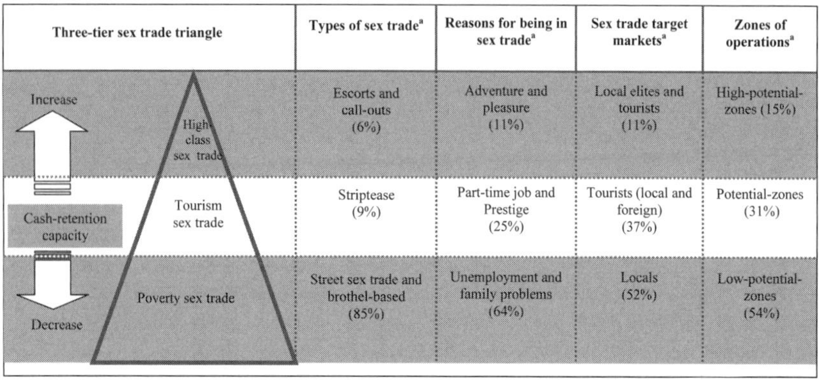

Three-tier sex trade triangle	Types of sex trade[a]	Reasons for being in sex trade[a]	Sex trade target markets[a]	Zones of operations[a]
Increase ↑ — High class sex trade	Escorts and call-outs (6%)	Adventure and pleasure (11%)	Local elites and tourists (11%)	High-potential-zones (15%)
Cash-retention capacity — Tourism sex trade	Striptease (9%)	Part-time job and Prestige (25%)	Tourists (local and foreign) (37%)	Potential-zones (31%)
Decrease ↓ — Poverty sex trade	Street sex trade and brothel-based (85%)	Unemployment and family problems (64%)	Locals (52%)	Low-potential-zones (54%)

Figure 10.1 Tourism-oriented sex trade in Kenya

Note: [a] Figures in parentheses are percentages in relation to the general sex business.

who promise them jobs outside their home areas/countries. While away from their homes, they are forced into sex trade. However, the issues of child sex tourism in Kenya, in Africa, and indeed worldwide, cannot be solved by a domestic solution alone. It requires internationally co-ordinated action. Furthermore, it will certainly be very difficult to intervene where family and other adults who occupy role-model positions are in collusion with human traffickers.

Gender discrimination also works in tandem with poverty – in many Kenyan societies, African societies, and in fact many societies in the developing countries, girls are not given equal education opportunities to boys (see, for example, Chapter 6). As a result, they have fewer employment opportunities in the formal sector than their male counterparts. The end result is that they must find other means of earning a living, including trading in sex.

Feminisation of poverty aside, this book has shown that most of the CSWs in Kenya operate within the poverty sex trade in the three-tier sex trade model (Figure 1.1). Therefore, Kenya's tourism-oriented sex trade can be summarised schematically as in Figure 10.1.

Figure 10.1 indicates that 70 per cent of the CSWs in Kenya transact their sex business in the streets. Fifty-two per cent of Kenyan sex workers get their business from the local markets, and 64 per cent of them are in sex trade courtesy of a combination of lack of employment opportunities (59 per cent) and family problems (5 per cent). In situations where attempts have been made to 'professionalise' the tourism-oriented sex trade, for example in Malindi Area (Figure 6.1), the majority of CSWs (54 per cent) operate in the low-potential zones, 31 per cent in the potential zones and 15 per cent in the high-potential zones. Low-potential zones are characterised by low revenue-generation, while high-potential zones are associated with comparatively higher incomes, as the CSWs serve well-to-do clients – local elites and tourists. Further, as one moves from low-potential to high-potential zones, the number of middlemen between CSWs and their clients reduces. Thus, sex workers in high-potential zones enjoy a high cash-retention capacity, and hence they fall under the high-class sex trade depicted in Figure 1.1.

The Marriage between Tourism and Sex Trade

Existing literature has shown that the marriage between sex trade and tourism (sex tourism) is a reality of international tourism, accounting for a sizeable portion of all tourism activities. In addition, this book has revealed that a linkage between tourism and sex trade in Kenya exists. Sex workers in Kenya's coastal region, a key tourism destination in the country, feel that their sex business depends on tourism. However, it would be inaccurate to consider sex trade as a creation of tourism alone, because there is a direct and clear analogy with sex trade in urban centres in other parts of the country, where CSWs are disproportionately drawn from disadvantaged sections of the population who may have similar economic and other hardships. Sex trade existed in the society long before the current levels of tourism, but probably never came close to today's scale.

Moreover, quite a number of tourists who consume commercial sexual services are often opportunistic users of the existing sex industry infrastructure at a destination. In any case, sex tourism would be impossible if there is no already established sex industry aimed at local residents (Bauer and McKercher 2003). It would thus be misleading to wholly blame the rise of a domestic sex industry on the parallel expansion of the tourism industry within a given destination. Tourists are only part of the overall market for commercial sexual services. In a simpler manner, as is the case with many tourism attractants, end sexual consumers are typically a mix of the visitors and the visited. While the visitors increase the amount of available cash, the visited stimulate the local economy through circulating this money. Subsequently, it stands to reason that as the tourism industry in Kenya expands, so will the sector of the local sex industry that services international tourists. Thus, by preparing for that 'development', rather than adopting the conventional knee-jerk, anti-sex trade reaction, Kenya as a tourism destination can support those CSWs who will work in that sector of its tourism industry. This will ultimately cushion the country from the usual foreign dominance and manipulation of the local sex industry that typifies organised sex tourism in most developing countries such as Cambodia, the Philippines, Thailand and Vietnam. This book has consequently shown that the nexus between sex trade and tourism in Kenya is far more multi-faceted than the literature commonly documents.

Again, the present study has revealed that tourism is clearly an element in the reason for sex trade in Kenya, but so is culture, the pattern of economic development, poverty and wealth distribution, and material interests (see, for example, Figure 3.4). Examining government reports on sex tourism often gives an impression that something has been done. More often than not, the government, while taking 'action' on sex trade, and to a larger extent sex tourism, fails to recognise that in its own tourism advertising, it also promotes the commodification and objectification of the sexual body. But of course, the portrayal of such bodies in hotels' and tour operators' brochures has nothing to do with sex trade in the country – or does it?

The Complexity of Tourism-oriented Sex Trade

Sex trade is a social phenomenon which must be addressed from a more conceptual societal macro-perspective. For this reason, prohibition or control or legalisation of the sex business should take the opinions and interests of the society concerned, including STCs, into consideration. The trade should first of all be examined as a symptom of various social problems, poverty being the major one – it is an end result of many societal problems. Any attempt to deal with it without dealing with the social conditions behind it will definitely be futile. Emphasis should therefore first be placed in the elimination of its structural causes.

Related to the foregoing, the analyses in this modest text have shown that the relationship between sex trade and tourism is a product of an interplay between diverse factors – political, social, cultural and economic (Figure 3.4). In addition, the picture of the tourism-oriented sex trade provided in this book illustrates the distinct behaviours and characteristics that are inherent among the CSWs. This has important implications for those in the field of tourism management, and probably also for the self-proclaimed 'moral custodians', where the trend is to assume that STCs form a homogenous group. In consequence, arbitrary bias or stubborn resistance to seeing its deep structure by segments of our societies only compounds the situation.

The existence of MCSWs offering male–male commercial sex services is one of the key findings of the current study. This is more so due to the fact that Kenyan society is profoundly homophobic. In relation to this finding, this study has also revealed that moralistic positions are not the best perspectives from which to examine the complexity of the relationship between the sex trade and the tourism industry. Moralistic approaches ultimately render analysts numb to the predictable realities of sex trade and fail to put the profession into its proper context. However, one cannot condone the relationship between the two industries anywhere in the world if it is exploitative in nature and if the parties involved wield unequal power, as in the case of child sex tourism. In line with the foregoing, it is critical that decisions and policies that are formulated with respect to sex trade in general are subject to serious and considered debate. Such unbiased debate will result in a forum with a pragmatic focus on finding negotiated, consensual solutions to sex trade-related ills. Moreover, the development of the sex trade in Kenya seems to have established and reinforced a network of gender conditions which may take many years to dismantle. Based on this finding, therefore, it must be reiterated that until the corrosive power of the linkage between tourism and sex trade is subjected to a critical macro-analysis, effective social policies in the tourism industry can scarcely be framed.

From Sex Trade to Small-scale Tourism Projects

The most critical observation of the current study is that a percentage of sex workers in Kenya are ready to leave the sex business if better options are available. The government should therefore strive to provide alternative economic and social support mechanisms for those who are 'forced' to turn to sex trade as a source of employment. From a sustainable tourism viewpoint, Kenya's government should set up and/or fund small-scale tourism projects. Such projects should be dispersed, low-density and organised by the local communities themselves, thus fostering more meaningful interactions between local residents and tourists. The emphasis on small-scale, locally owned and locally run projects is based on the anticipation that tourism will increase the multiplier effects within the host community, as well as reducing foreign exchange leakages.

Moreover, small-scale (tourism) projects will result in the promotion of indigenous potential, the considerate handling of locally existing resources and the co-ordination of socio-economic and ecological aspects, while taking into account long-term effects of the actions taken and corporate strategies adopted. Thus, this work has not only provided the Kenyan government with a way to combat unsustainable/mass tourism and its obvious devastating effects, but also puts forward a possible option for alleviating poverty in Kenya. It begs noting that poverty (in its strict definition) encourages unsustainable practices in order to seek quick returns to meet immediate needs. If such practices are underestimated or not considered, one of the destination's most vital features – its people and their cultures – will crumble. Local culture will then convert from a renewable to a non-renewable resource. Based on this study's findings, one can categorise sex trade as an 'unsustainable practice' which has the capacity to undermine any form of sustainable tourism development in the country.

Based on this observation, therefore, if effective social policies in tourism are to be formulated, it is essential that debate and ideas about sex trade in general be brought into the wider public sphere. This will enhance integration of the issues related to the trade, rather than continuing to ignore them. Similarly, decisions and policies that are formulated with respect to sex tourism need to be subjected to vigorous debate. In the case of sex tourism in Kenya, this has never happened to date. Importantly, in attempting to control tourism-oriented sex trade, governments and individuals of conservative political and religious leanings should desist from engaging in sensationalised puritanical witch-hunts, as has started to occur in the country. Such an approach tends to ignore the wider structural, socio-economic and political contexts within which sex tourism and sex trade in general operate. Furthermore, it should be appreciated that like tourism, sex trade in Kenya has substantial national socio-economic impacts. In addition, it is needless to belabour the fact that a proportion of the income generated through sex trade is derived from the tourism industry – through the trickle-down effect or simply a multiplier effect.

However, although sex trade-related activities undeniably offer a substantial income to a proportion of Kenya's population, two issues remain unclear:

1. Do sex workers constitute a marginal layer in the social fabric of emerging as well as more established tourism destinations in Kenya?
2. Or do CSWs form a creative and innovative force that rightly deserves to be called entrepreneurial?

Nevertheless, what is certain is that CSWs are not just bearers of social norms and values, but rational, strategising actors.

By the same token, the existence of a percentage of CSWs who would be unwilling to leave the sex trade for an alternative profession means it is necessary for policymakers to consider the inter-relationships between material interests and political power. Most of these Kenyans work for their own sustenance and that of their families. They all wish to achieve a better standard of living. Thus, the number of CSWs in Kenya will not decrease as long as the root cause is not addressed – poverty! Furthermore, contradictions embedded in some socio-cultural aspects of Kenyan society in general should be scrutinised. This society, for example, has been tolerant towards men's promiscuity as long as they take care of their families. On the other hand, women are divided into 'virtuous daughters and good wives' and 'whores'. Women in the first group are required to be loyal to their fathers and husbands later in life. Those who fall into the second category are shunned by mainstream society and treated as outcasts. They are thus segregated in some quarters for the sake of 'protecting' public morality. This kind of double standard allows some people to be blind to the fact that CSWs have families and sometimes work in the sex industry for their benefit.

Overall, this book has discussed many pertinent aspects of the relationship between sex trade and tourism in Kenya. The study has revealed that knowledge of the reasons for being in the trade is useful in understanding the range of activities and the multi-faceted nature of CSW–customer relationships. As a result, regular police crackdowns on CSWs can only serve as a temporary remedy, as it only addresses the symptoms, rather than tackling the predisposing factors. Such crackdowns should be re-evaluated, and indeed re-focused.

In closing, it merits noting that whereas a number of recent studies have added insights into the sex (trade)–tourism phenomenon, there are still areas that require more attention and research. Unfortunately, the subject is highly charged with emotions, social prejudices, and socio-cultural and political bias, especially in the African context. It is high time, therefore, that researchers recognise its full importance, and thus treat the subject with more academic seriousness.

Areas for Future Research

Importantly, this study identifies a number of factors that clearly exert significant influence on CSWs – issues that have not hitherto been explored in depth, and which clearly merit serious academic attention. First, only a limited number of segmentation variables were investigated in this study, and there appear to be some theoretically exciting and strategically useful relationships to be tested further. Specifically, such areas include the CSW–tourist relationship, the tourist industry's views of the sex trade, and local people's perception of the role tourism has played in the growth of sex trade in their area. Second, further research should be carried out to investigate whether or not increased income from tourism among Malindi's population either improves or degrades the economic position of CSWs in the low-potential zones.

Finally, and perhaps most controversially, a thorough examination is needed of the merits and demerits of an organised and possibly decriminalised sex trade in Kenya, particularly in the urban centres. Whether this proposal is adopted or not, however, this study may have challenged or even confirmed the preconceptions that some people have regarding the relationship between sex trade, and the sex business in particular, and the tourism industry in general. In the mean time, sex tourism remains the masked face of the Kenya's booming industry.

If, at this point, we have raised more questions than we have answered, then our goal has been achieved.

Epilogue

Why Interest in Sex Tourism Research?

Many readers of my scholarly works have asked me: what do you really know about tourism-related sex trade? Why are you so passionate about sex tourism so as to publish so many papers and now a book? So far I have answered the first question while the answer to the second one is given below.

I have been motivated in the enormous task of writing this book mainly by my observations in the 'field' as I worked as a Government Tourist Officer at the Kenyan coastal region. Thus, the book's journey started twelve years ago, 26 June 1997 at 07h50 (Kenya's time), was the turning point. This is the date and time when, on the way to my office, I met Asha [pseudonym], an eleven-year-old, pure and virginal-looking street girl. She stopped me in her daily escapade to beg for money. I gave her two Kenyan pounds and then invited her for breakfast in a restaurant a block away from my office. As we took our breakfast, Asha narrated how she was pushed onto Mombasa streets and eventually into the sex trade by abject poverty. At the tender age of ten years she had lost count of the number of men she had 'sexually entertained' for cash. She could however, vividly recall how she had painfully lost her virginity to an Irish tourist on the eve of her eighth birthday. Asha noted: "I could not imagine failing to get the 25 US$ he was willing to give me for a sexual service". Her story was both chilling and compelling. I heard hundreds of similar stories from both underage and adult Kenyans during my six month assignment as the Secretary to Kenya's Government Task Force on coastal tourism. To my amazement none of the senior government bureaucrats in the Ministry of Tourism would admit the fact that sex tourism is the masked face of the Kenya's booming tourism industry. As a result I made a pledge to myself that someday I would tell the world part of the real story behind Kenya's, and indeed, Africa's, tourism industry.

I was fascinated as I began to delve into social science research at the lack of scientifically coherent studies on Kenya's tourism-oriented sex trade and at an absence of empathy for African child sex workers left to fend for themselves by the cruel dictates of grinding poverty. These children indulge in the sex trade because they see no hope in their future. These factors both separately and cumulatively triggered my scholarly interest in sex trade-related research. This was the germ of *Sex Tourism in Africa: Kenya's Booming Industry*.

Bibliography

Ackermann, T. and Filter, C. (1994) *Die Frau nach Katalog*. Freiburg: Herder.
Afrol (2003) 'European involved in Gambian Child Sex Tourism'. *Afrol News*, <http://www.afrol.com>: 1–3 (accessed 16 May 2003).
Agrusa, J. (2003) 'AIDS and Tourism: A Deadly Combination'. In T. Bauer and B. McKercher (eds), *Sex and Tourism: Journeys of Romance, Love, and Lust*. New York: The Haworth Hospitality Press, pp. 167–80.
Akama, J. (1999) 'The Evolution of Tourism in Kenya'. *Journal of Sustainable Tourism*, 7(1): 6–25.
—— (2002) 'The Role of Government in the Development of Tourism in Kenya'. *International Journal of Tourism Research*, 4: 1–13.
Alleyne, D. and Boxill, I. (2003) 'The Impact of Crime on Tourist Arrivals in Jamaica'. *International Journal of Tourism Research*, 7: 99–135.
Archavanitkul, K. and Guest, P. (1994) 'Migration and the Commercial Sex Sector in Thailand'. *Health Transition Review*, 4 (supplement): 273–95.
AVERT.ORG (2004) *HIV and AIDS in Thailand*, <http.//www.avert.org/>: 1–6.
Bachmann, P. (1988) *Tourism in Kenya: A Basic Need for Whom?* Berne: Peter Lang.
Backwesegha, C. (1982) *Profiles of Urban Prostitution: A Case Study of Uganda*. Nairobi: Kenya Literature Bureau.
Baillie, W. (1980) 'International Travel Trends in Canada'. *Canadian Geographer*, 24(1): 13–21.
Bauer, T. and McKercher, B. (eds) (2003) *Sex and Tourism: Journeys of Romance, Love, and Lust*. New York: The Haworth Hospitality Press.
Boissevain, J. (1974) *Friends of Friends: Networks, Manipulators and Coalitions*. Oxford: Basil Blackwell.
Brenner, L. (2005) 'State-planned Tourism Destinations: The Case of Huatulco, Mexico'. *Tourism Geographies*, 7(2): 138–64.
Britton, S. (1982) 'The Political Economy of Tourism in the Third World'. *Annals of Tourism Research*, 9(3): 331–58.
Brown, F. and Hall, D. (eds) (2000) *Tourism in Peripheral Areas*. Clevedon: Channel View Publications.
Brown, L. (2000) *Sex Slaves: The Trafficking of Women in Asia*. London: Virago.
Burns, P. and Holden, A. (1995) *Tourism: A New Perspective*. London: Prentice-Hall.
Burton, K. (1995) 'Sex Tourism and Traditional Australian Male Identity'. In: M. Lanfant, J. Allcock and E. Bruner (eds), *International Tourism*. London: Sage Publications, pp. 194–211.

Butler, R. (1980) 'The Concept of a Tourist Area Life Cycle of Evolution: Implications for Management of Resources'. *Canadian Geographer*, 14(1): 5–12.

Cabezas, A. (1999) 'Women's Work is Never Done: Sex Tourism in Sosua, the Dominican Republic'. In: K. Kempadoo (ed.), *Sun, Sex and Gold: Tourism and Sex Work in the Caribbean*. Lanham, MD: Rowman & Littlefield, pp. 93–124.

Caplan, G. (1984) 'The Facts of Life about Teenage Prostitution'. *Crime & Delinquency*, 20(1): 69–74.

Cazes, G. (1992) *Tourisme et tiers-monde, un bilan contraversé: les nouvelles colonies de vacances*. Paris: L'Harmattan.

Chapman, C. (2005) 'If You Don't Take a Job as a Prostitute, We Can Stop Your Benefits. *News-Telegraph*, <http://www.telegraph.co.uk>: 1–2 (accessed 21 January 2005).

Cohen, E. (1982) 'Thailand Girls and *Farang* Men: The Edge of Ambiguity'. *Annals of Tourism Research*, 19(3): 403–28.

—— (1988a) 'Authenticity and Commoditization in Tourism'. *Annals of Tourism Research*, 15(3): 371–86.

—— (1988b) 'Tourism and Aids in Thailand'. *Annals of Tourism Research*, 15: 467–86.

—— (1989) 'Primitive and Remote: Hill Tribe Trekking in Thailand'. *Annals of Tourism Research*, 16: 30–61.

—— (1993) 'Open-ended Prostitution as a Skilful Game of Luck: Opportunity Risk and Security among Tourist-oriented Prostitutes in a Bangkok Soi'. In: M. Hitchcock, V. King and M. Parnwell (eds), *Tourism in South-East Asia*. London: Routledge, pp. 155–78.

Corbin, A. (1990) *Women for Hire: Prostitution and Sexuality in France after 1850*. Cambridge, MA: Harvard University Press.

Cuvelier, P. (1998) *Anciennes et nouvelles formes de tourisme: une approche socio-économique*. Paris: L'Harmattan.

Dahles, H. (1999) 'Tourism and Small Enterprises in Developing Countries: A Theoretical Perspective'. In H. Dahles and K. Bras (eds), *Tourism and Small Entrepreneurs: Development, National Policy and Entrepreneurial Culture – Indonesian Cases*. Elmsford, NY: Cognizant Communication Corporation, pp. 1–19.

Dahles, H. and Bras, K. (eds) (1999) *Tourism and Small Entrepreneurs: Development, National Policy and Entrepreneurial Culture – Indonesian Cases*. Elmsford, NY: Cognizant Communication Corporation.

De Kadt, E. (ed.) (1979) *Tourisme: passeport pour le développement? Regards sur les effets culturels et sociaux du tourisme dans les pays en développement*. Paris: Economica.

Delacoste, F. and Alexander, P. (eds) (1988) *Sex Work: Writings by Women in the Sex Industry*. London: Virago Press.

Dewailly, J.-M. (1999) 'Sustainable Tourist Space: From Reality to Virtual Reality?' *Tourism Geographies*, 1(1): 41–55.

Dieke, P. (1991) 'Policies for Tourism Development in Kenya'. *Annals of Tourism Research*, 19: 69–90

Enberg, T. (2008) 'You've Come a Long Way, Baby'. *24 Hours* (Ottawa: Sun Media Corporation), 27 November: 16.

End Child Prostitution, Pornography and Trafficking (ECPAT) (2002) 'Sustainable Tourism Development and the Protection of Child from Sexual Exploitation: New Research to Shed light on Child Sex Tourism in The Gambia'. *ECPAT International Newsletter*, 41 (October): 1.

Fanon, F. (1966) *The Wretched of the Earth*. New York: Grove Press.

Findley, S. and Williams, L. (1991) *Women Who Go and Women Who Stay: Reflections of Family Migration Processes in a Changing World.* Working Paper 176, Geneva: International Labour Organisation.

Finnegan, F. (1979) *Poverty and Prostitution: A Study of Victorian Prostitutes in York*. Cambridge: Cambridge University Press.

Gallagher, C. and Laquer, T. (eds) (1987) *The Making of the Modern Body: Sexuality and Society in the 19th Century*. Berkeley, CA: University of California Press.

Gitonga, A. and Anyangu, S. (2008) 'More than 1,000 Women Trafficked to Germany'. *East African Standard* (Standard Newspapers, <http://www.eastandard.net/>): II (accessed 20 November 2008).

Gould, S. (1995) 'Sexualized Aspects of Consumer Behavior: An Empirical Investigation of Consumer Lovemaps'. *Psychology & Marketing*, 1(2): 395–413.

Graburn, N. (1979) 'Tourism: The Sacred Journey'. In: V. Smith (ed) *Hosts and Guests: The Anthropology of Tourism*. Philadelphia, PA: University of Pennsylvania Press, pp. 76–87.

—— (1983) 'Tourism and Prostitution'. *Annals of Tourism Research*, 10(3): 437–42.

Günther, A. (1998) 'Sex Tourism Without Sex Tourists'. In M. Oppermann (ed.), *Sex Tourism and Prostitution: Aspects of Leisure, Recreation, and Work*. Elmsford, NY: Cognizant Communication Corporation, pp. 71–80.

Gursoy, D., Jurowski, C. and Uysal, M. (2001) 'Resident Attitudes: A Structural Modelling Approach'. *Annals of Tourism Research*, 29(1): 79–105.

Hall, M. (1994) *Tourism and Politics: Policy, Power and Place.* Chichester: John Wiley.

—— (1996) 'Gender and Economic Interests in Tourism Prostitution: The Nature, Development and Implications of Sex Tourism in South-East Asia'. In Y. Apostolopoulos, S. Leivadi and A. Yiannakis (eds), *The Sociology of Tourism: Theoretical and Empirical Investigations.* London: Routledge, pp. 265–80.

Hall, M. and Tucker, H. (2004) *Tourism and Postcolonialism: Contested Discourses, Identities and Representations*. New York: Routledge.

Hanmer, J., Radford, J. and Stanko, E. (1989) *Women, Policing and Male Violence*. London: Routledge.
Harrell-Bond, B. (1978) *A Window on an Outside World: Tourism as Development in the Gambia*. American Universities Field Staff Report 19. Hanover, NH: American Universities Field Staff.
Harrison, D. (1994) 'Tourism and Prostitution: Sleeping with the Enemy? The Case of Swaziland'. *Tourism Management*, 15(6): 455–43.
Hobson, B. (1990) *Uneasy Virtue: The Politics of Prostitution and the American Reform Tradition*. Chicago, IL: University of Chicago Press.
Holloway, C. (1994) *The Business of Tourism*, 4th edn. London: Pitman Publishing.
IUCN (1991) *Protected Areas of the World: A Review of National Systems*, vol. 3: *Afrotropical*. Gland, Switzerland: World Conservation Union.
Honey, M. (1999) *Ecotourism and Sustainable Development: Who Owns Paradise?* Washington, DC: Island Press.
Jacobs, J. (1984) *Cities and the Wealth of Nations: Principles of Economic Life*. New York: Random House.
Jackson, T. (1986) *The Laws of Kenya: Cases and Statutes*. Nairobi: Kenya Literature Bureau.
Jago, L. (2003) 'Sex Tourism: An Accommodation Provider's Perspective'. In: T. Bauer and B. McKercher (eds) *Sex and Tourism: Journeys of Romance, Love, and Lust*. New York: The Haworth Hospitality Press, pp. 85–154.
Jallow, E. (2004) 'The State of Gambian Children'. *Online Newsdesk*, <http://www.ecpat.net>: 1–5 (accessed 6 December 2004).
Jenkins, C. and Henry, B. (1982) 'Government Involvement in Tourism in Developing Countries'. *Annals of Tourism Research*, 9: 499–521.
Kempadoo, K. (ed.) (1999) *Sun, Sex and Gold: Tourism and Sex Work in the Caribbean*. Lanham, MD: Rowman & Littlefield.
Kelly, L. (1988) *Surviving Sexual Violence*. Cambridge: Polity Press.
Kenya, Government of (1967) *Economic Survey*. Nairobi: Government Printers.
—— (1987) *Economic Survey*. Nairobi: Government Printers.
—— (1989) *Economic Survey*. Nairobi: Government Printers.
—— (1990) *Economic Survey*. Nairobi: Government Printers.
—— (1991) *Economic Survey*. Nairobi: Government Printers.
—— (1996) *Economic Survey*. Nairobi: Government Printers.
—— (1998) *Economic Survey*. Nairobi: Government Printers.
—— (2000) *Economic Survey*. Nairobi: Government Printers.
—— (2001) *Economic Survey*. Nairobi: Government Printers.
—— (2002a) *Economic Survey*. Nairobi: Government Printers.
—— (2002b) *Malindi District Development Plan: 2002–2008*. Nairobi: Government Printers.
—— (2003) *Economic Survey*. Nairobi: Government Printers.
—— (2004) *Economic Survey*. Nairobi: Government Printers.
—— (2006) *Economic Survey*. Nairobi: Government Printers.

Kenya Wildlife Service (KWS) (1990) *A Policy Framework and Development Program, 1991–1996*, Nairobi: Government Printers.
—— (1994) *Wildlife–human Conflicts in Kenya*. Nairobi: Government Printers.
—— (1995) *Tourism Development Policy and Pricing Study: A Survey of Visitors to Selected KWS Parks*, Working Paper 7. Nairobi: Government Printers.
—— (1997). *Maintaining Bio-diversity into the 21st Century*. Nairobi: Government Printers.
Kibicho, W. (2003) 'Community Tourism: A Lesson from Kenya's Coastal Region'. *Journal of Vacation Marketing*, 10(1): 33–42.
—— (2004a) 'A Critical Evaluation of How Tourism Influences the Operations of Commercial Sex Workers'. *Annals of Leisure Research*, 7(3/4): 188–201.
—— (2004b) 'Tourism and the Sex Trade: Roles Male Sex Workers Play in Malindi, Kenya'. *Tourism International Review*, 7(3/4): 129–41.
—— (2005a) 'Impacts of Tourism in Malindi: An Analysis of Gender Differences in Perception'. *ASEAN Journal on Tourism and Hospitality*, 4(1): 83–96.
—— (2005b) 'Tourism and the Sex Trade in Kenya's Coastal Region'. *Journal of Sustainable Tourism*, 13(3): 256–80.
—— (2005c) 'Tourisme et parcs nationaux au Kenya: Ville contre la société rurale locale?' (unpublished doctoral thesis). Lyon: Université Lumière Lyon 2.
—— (2006a) 'An Evaluation of Critical Factors to Successful Community-based Tourism in Kenya'. *Cahiers de Tourisme & Urbanisme de Lyon*, 50(112/113): 106–37
—— (2006b) 'An Evaluation of Critical Factors to Successful Community-based Tourism in Kenya'. *Journal of Hospitality and Tourism*, 4(1): 137–53.
—— (2006c) 'Impacts socioculturels du tourisme: le cas de Malindi, Kenya'. *Cahiers de Tourisme & Urbanisme de Lyon*, 51(112/113): 44–65.
—— (2006d) 'Service Quality in Malindi's Tourism Industry: A Study of Tourist Perceptions'. *ASEAN Journal of Hospitality & Tourism Research*, 12(3): 218–31.
—— (2006e) 'Tourists to Amboseli National Park: A Factor-cluster Segmentation Analysis'. *Journal of Vacational Marketing*, 12(3): 218–31.
—— (2007) *Tourisme au Pays Maasaï: de la destruction au développement durable*. Paris: L'Harmattan.
Kithaka, G. (2004) 'Confessions of a Former Twilight Girl'. *Daily Nation* (Nation Group, <http://www.nationmedia.com>): I–III (accessed 4 August 2004).
Kohm, S. and Selwood, J. (1998) 'The Virtual Tourist and Sex in Cyberspace'. In: M. Opperman (ed.) *Sex Tourism and Prostitution: Aspects of Leisure, Recreation and Work*. Elmsford, NY: Cognizant Communication Corporation, pp. 123–31.
Krippendorf, J. (1977) *Les dévoreurs de paysages: le tourisme doit-il détruire les sites qui le fait?* Lausanne: Editions 24 Heures.
Launer, E. (1993) *Zum Beispiel Sextourismus*. Göttingen: Lamnv.
Leheny, D. (1995) 'A Political Economy of Asian Sex Tourism'. *Annals of Tourism Research*, 2(2): 367–84.

Lickorish, L. Jefferson, A. Bodlender, J. and C. Jenkins, C. (1991) *Developing Tourism Destinations: Policies and Perspectives*. Harlow: Longman.

Maslow, A. (1954) *Motivation and Personality*. New York: Harper.

Mathews, H. (1978) *International Tourism: A Political and Social Analysis*. Cambridge: Shenkman Publishing.

Mathieson, A. and Wall, G. (1982) *Tourism: Economic, Physical and Social Impacts*. Harlow: Longman.

Maurer, M. (1991) *Tourismus, Prostitution, AIDS*. Zurich: Rotpunktverlag.

McLeod, E. (1982) *Women Working: Prostitution Now*. London: Croom Helm.

McLintock, A. (1992) 'Gonad the Barbarian and the Venus FlyTrap: Portraying the Female and Male Orgasm'. In L. Segal and M. McIntosh (eds), *Sex Exposed: Sexuality and the Pornography Debate*. London: Virago Press, pp. 101–12.

Medlik, S. (1991) *Managing Tourism*. Oxford: Butterworth/Heinemann.

Mensah, J. (2005) *Black Canadians: History, Experiences, Social Conditions*. Winnipeg: Fernwood Publishing.

Middleton, S. (1993) *Educating Feminist: Life Histories and Pedagogy*. New York: Teachers College Press.

Mingmong, S. (1981) 'Official Blessings for the Brothel of Asia'. *Southeast Asia Chronicle*, 7(8): 24–5.

Mulhall, B., Hu, M., Thompson, M., Lin, F., Lupton, D., Mills, D., Maund, M., Cass, R. and Miller, D. (1994) 'Planned Sexual Behavior of Young Australian Visitors in Thailand'. *The Medical Journal of Australia*, 15(8): 530–35.

Mungai, M. (1998) 'Abstaining from Sex Causes No Disease: On Aids against AIDS'. *Daily Nation* (Nairobi), 24 December: II.

Muroi, H. and Sasaki, N. (1997) *Tourism and Prostitution in Japan*. In T. Sinclair (ed.), *Gender, Work and Tourism*. London: Routledge, pp. 181–219.

Mwakisha, J. (1995) 'Trade in the Flesh: The Kenya Link'. *Daily Nation* (Nairobi), 4 October: I.

Mwangi, W. (1995) 'Ban Homosexuality'. *East African Standards* (Nairobi), 16 July: 3.

Naibavu, T. and Schutz, B. (1974) 'Prostitution: Problem or Profitable Industry?'. *Pacific Perspective*, 3(1): 39–68.

Ndune, K. (1996) 'Position Paper on the Commercial Exploitation of Children. Ministry of Tourism and Wildlife'. Mombasa: unpublished paper.

O'Grady, R. (1992) *The Child and the Tourists*. Bangkok: ECPAT.

—— (1994) *The Rape of the Innocent*. Bangkok: ECPAT.

Oppermann, M. (1998) 'Introduction'. In: M. Oppermann (ed.), *Sex Tourism and Prostitution: Aspects of Leisure, Recreation, and Work*. Elmsford, NY: Cognizant Communication Corporation, pp. 1–19.

Oppermann, M., McKinley, S. and Chon, K. (1998) 'Marketing Sex and Tourism Destinations'. In: M. Oppermann (ed.), *Sex Tourism and Prostitution: Aspects of Leisure, Recreation, and Work*. Elmsford, NY: Cognizant Communication Corporation, pp. 12–22.

Peake, R. (1989) 'Tourism and Swahili Identity in Malindi Old Town, Kenyan Coast'. *Africa*, 59(2): 209–20.

Phillip, J. and Dann, G. (1998) 'Bar Girls in Central Bangkok: Prostitution as Entrepreneurship'. In: M. Oppermann (ed.), *Sex Tourism and Prostitution: Aspects of Leisure, Recreation, and Work*. Elmsford, NY: Cognizant Communication Corporation, pp. 60–70.

Phongpaichit, P. (1981) 'Bangkok Masseuses: Holding Up the Family Sky'. *Southeast Asia Chronicle*, 7(8): 15–23.

Pickering, H. and Wilkins, H. (1993) 'Do Unmarried Women in African Towns Have to Sell Sex, or is it a Matter of Choice?'. *Health Transition Review*, 3 (supplement): 17–27.

Plant, M. (ed.) (1990) *AIDS, Drugs and Prostitution*. London: Routledge.

Provincial Tourism Liaison Committee (PTLC) (1998) 'A Report by the Task Force on Beach Operations'. Mombasa: unpublished government report.

Pruitt, D. and LaFont, S. (1995) 'For Love and Money: Romance Tourism in Jamaica'. *Annals of Tourism Research*, 22(2): 367–84.

Rao, N. (2003) 'The Dark Side of Tourism and Sexuality: Trafficking of Nepali Girls for Indian Brothels'. In: T. Bauer and B. McKercher (eds), *Sex and Tourism: Journeys of Romance, Love, and Lust*. New York: The Haworth Hospitality Press, pp. 155–66.

Robinson, G. (1989) 'AIDS fear triggers Thai action'. *Asia Travel Trade*, 21: 11.

Rogers, J. (1989) 'Clear Links: Tourism and Child Prostitution'. *Contours*, 4(2): 20–22.

Rutherford, J. (1997) *Forever England: Reflections on Masculinity and Empire*. London: Lawrence & Wishart.

Ryan, C. and Hall, M. (2001) *Sex Tourism: Marginal People and Liminalities*. London: Routledge.

Ryan, C., and Kinder, R. (1996) 'Sex, Tourism and Sex Tourism: Fulfilling Similar Needs?'. *Tourism Management*, 17(7): 507–18.

Ryan, J. (1997) *Picturing Empire: Photography and the Visualization of the British Empire*. London: Reaktion Books.

Runte, A. (1987) *National Parks: The American Experience*. London: University of Nebraska Press.

Sarpong, J. (2002) *Perspective of Issues of Trafficking of Women and Children in West Africa*. Ghana: African Women Lawyers Association.

Scott, J. (2000) 'Peripheries, Artificial Peripheries and Centres. In: F. Brown and D. Hall (eds), *Tourism in Peripheral Areas*. Clevedon: Channel View Publications, pp. 58–73.

Seabrook, J. (1996) *Travels in the Skin Trade: Tourism and the Sex Industry*. London: Pluto Press.

Sinclair, M. (1990) *Tourism Development in Kenya*. Washington, DC: World Bank.

Sindiga, I. (1995) 'The Dark Side of Tourism in Kenya with Special Reference to the Coast'. Eldoret, Kenya: MOI University, Department of Tourism, unpublished paper.

—— (1999) *Tourism and African Development: Change and Challenge of Tourism in Kenya*. Aldershot: Ashgate.

Smart, C. (1992) *Regulating Womanhood: Historical Essays on Marriage, Motherhood and Sexuality*. London: Routledge.

Smith, M. and Duffy, R. (2003) *The Ethics of Tourism Development*. London: Routledge.

Smith, V. and Eadington, W. (1992) *Tourism Alternatives*. New York: John Wiley.

Stanley, L. (ed.) (1990) *Feminist Praxis*. London: Routledge.

Symanski, R. (1981) *The Immoral Landscape: Female Prostitution in Western Societies*. Toronto: Butterworth.

TAT News (2004) 'Thailand in the Global Fight against the Commercial Sexual Exploitation of Children (CSEC)'. <http.//www.tatnews.org/>: 1–4 (accessed 17 May).

Thomas, K. (1983) *Man and the Natural World: Changing Attitudes in England (1500–1800)*. Harmondsworth: Penguin.

Thompson, W. and Harred, J. (1992) 'Topless Dancers: Managing Stigma in a Deviant Occupation'. *Deviant Behaviour*, 13(2): 291–311.

Thoya, F. (1998) 'Tourism: 200,000 May be Retrenched'. *Sunday Nation* (Nairobi), 3 May: 1–3.

Trovato, B. (2004) 'A Look at Sex Tourism in Gambia, where Euroblimps Waddle to be Coddled by the Local Lads'. *Cape Times*, <http://www.capetimes.co.za>: 1–2 (accessed 17 May 2004).

Truong, T. (1983). 'The Dynamics of Sex Tourism: The Case of Southeast Asia?'. *Development and Change*, 14(4): 533–53.

Turner, L. and Ash, J. (1975) *The Golden Hordes: International Tourism and the Pleasure Periphery*. London: Constable.

Walkowitz, J. (1980) *Prostitution and Victorian Society*. Cambridge: Cambridge University Press.

Walsh, J., Jamorzy, U. and Burr, S. (2001) 'Sense of Place as a Component of Sustainable Tourism Marketing'. In: F. McCool and R. Moisey (eds), *Tourism, Recreation and Sustainability: Linking Culture and the Environment*. Wallingford: CABI Publishing, pp. 195–216.

Watson, G. and Kopachersky, J. (1996) 'Interpretations of Tourism as a Commodity'. In: Y. Apostolopoulos, S. Leivadi and A. Yiannakis (eds), *The Sociology of Tourism: Theoretical and Empirical Investigations*. London: Routledge, pp. 281–97.

Weaver, D. (1998) *Ecotourism in the Less Developed World*. Wallingford: CABI Publishing.

Wels, H. (2000) *Fighting Over Fences: Organisational Co-operation and Reciprocal Exchange Between the Save Valley Conservancy and its Neighbouring Communities*. Amsterdam: Breukelen.

Wilson, D. (1997). 'Paradoxes of Tourism in Gao'. *Annals of Tourism Research*, 24(2): 52–7.
World Commission on Environment and Development (WCED) (1987) *Our Common Future*. Oxford: Oxford University Press.
World Tourism Organisation (WTO) (2004) *Indicators of Sustainable Tourism: A Guidebook*. Madrid: WTO.

Index

Page numbers in *italics* refer to figures.

abolitionists vs regulationist debate 44–8
adventurism as motivation 107, 136–8
advertising
    commercial sexual services 93–5
    see also marketing
Africa
    abolitionists vs regulationist debate 46–7
    attitudes to sex trade 2–3
    context 79–80
    traditional society 111–12, 117–18, 131–3, 138–9, 153–4, 170, 212
Agrusa, J. 160
AIDS *see* HIV-AIDS; sexually transmitted diseases (STDs)
anonymity of sex tourists 105–6
Australia 51

Backwesegha, C. 3, 23, 34, 64, 76, 117, 131, 138, 140
Baillie, W. 203, 204
bars 92, 93
battered women 131–2
Bauer, T. and McKercher, B. 1, 29, 31, 34, 80, 87, 88, 89, 144, 209
beach boys 116, 117–18, 137, 162
body massage 92
Boissevain, J. 34, 38, 174, 175
bonded commercial sex workers (CSWs) 84–5
brothel-based sex trade 84–5, 91

Cabezas, A. 150, 192
call-out sex trade 89–90, 91
Chapman, C. 44, 147
cheapness as motivation of sex tourists 81–2, 83, 105
child sex tourism 19–21
    international legal responses to 50–2
    personal accounts 113, 114–15, 146
    and poverty, relationship between 112–15
childbirth and pregnancy 132–3, 134, 138
Children's Act (2002), Kenya 50–1
coastal region, Kenya 121–41, 143–56
    see also Malindi
Cohen, E. 1, 3, 5, 13, 18, 22, 30, 83, 88, 100–1, 124, 135, 137, 140, 143, 144, 152, 154, 162, 163, 167, 169, 193
colonialism
    and sex worker–tourist relationship 134–5
    tourism development, pre-independent Kenya 62–4, 75
commercial sex workers (CSWs) 22–4
    bonded 84–5
    personal accounts 46–7, 87, 90, 94, 95, 126–38 *passim*, 151–6 *passim*, 172, 184, 185, 186, 192
    children 113, 114–15, 146
    male 137, 168–9, 174
'commission' paid to police by CSWs 151
community attitudes 37–8
community participation 199–201
condom usage 144–6
Corbin, A. 26, 29, 49
core-periphery economic analysis 159
corruption
    'commission' paid to police by CSWs 151
    'protection fee' paid by strip club owners 87–8
Criminal Law (Amendment) Act (2003), Kenya 51
criminalisation *see under* legal aspects
CSWs *see* commercial sex workers

Dahles, H. 117, 118–19, 167, 175, 176
definitions and concepts 14–27, 30–1

developing countries 18, 105
discothèques 92, 93

earnings 18, 189–92
    and financial negotiations 81–2, 83, 168–9
economic issues *see* poverty; tourism development, Kenya
employment
    alternative strategies 37, 140–1, 211–12
    part-time 129–31
    unemployment 127–9
entrepreneurs, male sex workers (MCSWs) as 174–7
escort sex trade 89–90, 91
external and internal environment 35

family problems
    and poverty, relationship between 133–4
    pregnancy and childbirth 132–3, 134, 138
    and responsibilities as motivation of CSWs 131–3, 153–4
female romance tourists 101–3
female sex tourists 101–4, 167–9
    personal accounts 103, 105, 106, 107, 108, 117
female strippers 86–7
female students 89–90

gay sexual services 170–1, 210
gay strip clubs 88
gender relations/inequalities 111–12, 117–19, 212
    see also family problems
Germany 44, 49, 147–8
Gitonga, A. and Anyangu, S. 26, 27, 35, 201
Gould, S. 204, 205
Graburn, N. 17, 85, 143, 162
Günther, A. 1, 19, 53, 83, 107, 110

Hall, M. 17, 66, 68–9, 74, 85
    Ryan, C. and 3, 5, 17, 44, 79–80, 166, 169, 171
    and Tucker, H. 107

harassment by law enforcement agents 150–1
health risks *see* sexually transmitted diseases
high-class sex trade 33
historical perspective
    of sex tourism 27–30
    legal aspects 42–4
    of tourism development, Kenya 59–61
HIV–AIDS 148–9, 165, 194–5
    see also sexually transmitted diseases (STDs)
'hot-pillow trade' 93
hotel workers 122–3, 130, 131
human trafficking, Kenya 25–7

impulse consumption as motivation of sex tourists 107–8
income *see* earnings
information technology *see* Internet/World Wide Web
internal and external environment 35
Internet/World Wide Web 18–19, 95, 108–9
    virtual reality as alternative to sex tourism 109–11
Italy 51

Jackson, T. 170
Jago, L. 1, 43, 46, 48, 49, 87, 88, 89, 166
Japanese sex tourism 29, 47–8, 98–9

Kenya 79–95, 207–12
    coastal region 121–41, 143–56
    human trafficking 25–7
    legal aspects 52–4, 104–5, 170
    Ministry of Tourism and Wildlife 65–7, 74
    ratification of UN Convention on the Rights of the Child 50–1
    socio-economic conditions 121–41
    sustainable tourist development 195–9
    types of commercial sexual services 91–3
    types of tourism-oriented sex tourism 84–91
    see also Malindi; tourism development, Kenya

Kenya Tourist Development Corporation (KTDC) 57, 65, 68, 74
Kenyan Tourist Board (KTB) 74–5, 77
Kenyan Wildlife Service (KWS) 69, 70, 71
Kibicho, W. 2, 4, 5, 22, 25, 30, 31, 44, 45, 47, 52, 55, 60, 66, 67, 68, 69, 71, 72, 73, 74, 77, 80, 83, 84, 85, 87, 89, 98, 103, 111, 116, 121, 127–8, 133–5, 138, 147, 149, 151, 157, 161–2, 164, 169, 170, 176, 189, 197, 200, 202, 205
King, Martin Luther 107
Kohm, S. and Selwood, J. 109, 110, 204–5

law enforcement agents, harassment by 150–1
legal aspects 37, 41–2, 173
  abolitionists vs regulationist debate 44–8
  criminalisation vs decriminalisation 54–5
  criminalisation vs regulation 49–52
  historical and international perspectives 42–4
  homosexuality 170
  international responses to child sex tourism 50–2
  Kenya 52–4, 104–5, 170
  and motivations of sex tourists 104–5
Leheny, D. 124, 144, 203
Lickorish, L. et al. 72, 74

McLintock, A. 107, 154
magazine advertisments 94, 95
male sex tourists 99–101
  personal accounts 20, 81–2, 100, 145, 190–1
male sex workers (MCSWs)
  beach boys 116, 117–18, 137, 162
  as entrepreneurs 174–7
  personal accounts 137, 168–9, 174
  Rastafarian 117–18
  research 162
  services offered 166–74
  and tourism, link between 163–6
male strippers 87–8
male–female sexual services by male sex workers 167–9

male–male sexual services 170–1, 210
Malindi 179–206
  tourist development in 158–62
  see also coastal region, Kenya
Malindi Welfare Association (MWA) 4, 136
  male sex workers 163
  organisational structure 181–2
  sex trade operation zones 183–5, 186–7, *188*, 189
  as source of social capital 185–9
marketing
  tourism and sex tourism 203–5
  see also advertising
mass tourism 17
  Kenya 77
Mensah, J. 135
methodological critique 3–5
Ministry of Tourism and Wildlife, Kenya 65–7, 74
mistreatment of CSWs by clients 151–2
motivations
  of CSWs 122–40
  of sex tourists 104–8
Muroi, H. and Sasaki, N. 29, 47, 98–9
Mwangi, W. 117, 170

Naibavu, T. and Schutz, B. 21, 150, 195
national AIDS awareness campaigns 148–9, 165
nationalities of female sex tourists 167–8
Ndune, K. 2, 19, 30, 63–4, 76, 77, 80, 99, 112, 118, 124, 125, 131, 133, 138, 147, 151
Netherlands 49, 147–8
New Zealand 49
newsagents' cards 94
newspaper advertisments 94
non-sexual services 25

operation areas 140
operation zones 183–5, 186–7, *188*, 189, 208–9
Oppenmann, M. 17, 19, 22, 109, 110, 112, 203, 204–5
'Others', stereotypical beliefs about 106–7, 117

part-time employment 129–31
peer abuse 156
penetrative sexual services 92–3
permanent clients 191
pimping services 171–4
pleasure motivation of CSWs 135–6
positive and negative aspects
    of sex tourism 192–4, 195, *196*
    of tourism development 16
poverty 22–3, 24, 31, 207–8
    and child sex tourism, relationship between 112–15
    and family problems, relationship between 133–4
    and socio-economic conditions, Kenya 121–41
    and sustainable tourism 201–2
    tourism role in alleviation 35–7
pregnancy and childbirth 132–3, 134, 138
prestige of sex worker–tourist relationship 134–5
private apartment-based sex trade 84–5, 91
product sexualisation 204
'protection fee' paid by strip club owners 87–8
Pruitt, D. and LaFont, S. 22, 99, 103, 117, 162, 195

Rao, N. 13, 167
Rastafarian men 117–18
recruitment of CSWs 26–7, 172–3
red light districts (RLDs) 24–5
regular encounters 192
regulation of sex trade *see under* legal aspects
Robinson, G. 195
rural vs rural economy, Malindi 158–61
Rutherford, J. 63
Ryan, C.
    and Hall, M. 3, 5, 17, 44, 79–80, 166, 169, 171
    and Kinder, R. 25, 203, 204

seasonality of tourist trade 122–5
sex tourism
    contextualisation 207–9
    definitions and concepts 14–27, 30–1
    historical background 27–30

positive and negative aspects 192–4, 195, *196*
spectrum *32*
sex tourists 21–2
    motivations of 104–8
    see also female sex tourists; male sex tourists
sex workers *see* commercial sex workers (CSWs); male sex workers (MCSWs)
sexual violence 151–2
sexually transmitted diseases (STDs) 143–9, 165, 194–5
Sindiga, I. 2, 20, 27, 30, 59, 63–4, 76, 80, 118, 125, 143, 151
small-scale business
    as alternative to sex trade 140–1, 211–12
    to mega-scale tourism, Kenya 67–8
social capital, Malindi Welfare Association (MWA) as source of 185–9
social change, role of tourism 36
social equity, community participation as 199–201
social stigmatisation of CSWs 153–5
soft-sell techniques 191–2
stereotypical beliefs about 'Others' 106–7, 117
stickers on street light posts 94
street sex trade 85, 91
striptease
    Kenya 85–9, 91
    US 43
Structural Adjustment Programme (SAP), Kenya 68
sustainable tourist development 194–205
systems approach 38

tart cards 94
teenage pregnancy 134
tourism
    definitions and concepts 14–17
    insecurity and destination sustainability 202–3
    seasonality 122–5
    and sex tourism, relationship between 2–3, 32–3, 35–8, 67–8, 72, 75–8,

80–1, 84–91, 97–9, 155–6, 160–1, 163–6, 173–4, 209–10
tourism development
  Kenya 57–9
    consolidation stage: 1997–2005 73–5, 76, 77
    development stage: 1979–96 68–72, 77
    exploration stage: pre-independence period 62–4, 75
    framework for analysis 61–77
    and growth of sex trade 67–8, 72, 75–8
    historic background and economic issues 59–61
    involvement stage: 1963–78 64–8, 75–7
    Malindi 158–62
  positive and negative aspects 16
  sustainable 194–205
traditional African society 111–12, 117–18, 131–3, 138–9, 153–4, 170, 212
transaction models 33–5

unemployment 127–9

United Nations (UN)
  Convention on the Rights of the Child 50–1
  Convention on Suppression of the Traffic in Persons and Exploitation of CSWs 26
United States (US) 43, 49, 51
urban migration 161–2
urban vs rural economy, Malindi 158–61

violence 155, 156, 187
  domestic 131–2
  sexual 151–2
virginity 172
virtual reality, as alternative to sex tourism 109–11
vocabulary 13–14

Wels, H. 63, 135
Wildlife Conservation and Management Department (WCMD), Kenya 70–1
World Tourism Organisation (WTO) 1, 17, 197, 199, 200, 202
World Wide Web *see* Internet/World Wide Web